# SOUTH AFRICAN POULTRY DISHES

# SOUTH AFRICAN POULTRY DISHES

Magdaleen van Wyk • Pat Barton

HUMAN & ROUSSEAU
Cape Town Pretoria

*Stylist Colleen Schwager-Poultney wishes to thank Magdaleen van Wyk for making her lovely home and its contents available to prepare and photograph the dishes illustrated here; Binnehuis Interiors in Cape Town and Stellenbosch for the loan of some beautiful accessories; Andreas van Wyk and Walter Köppe for their masterly touches to the paella, grilled chicken, Chinese chicken and rösti respectively, and Die Koskas in Stellenbosch for providing the fresh herbs, salads, vegetables and flowers.*

*Cover: Chicken with spiced fruit (p. 85)*


First published in 1989 by Human & Rousseau (Pty.) Ltd.
State House, 3-9 Rose Street, Cape Town
Atrium Building, 60 Glenwood Road, Pretoria
Cover Photograph by Cornel de Kock and styling by
Vivien MacLennan
Typography and cover design by Lizé Miller
Photography by David Briers
Illustrations by Rentia Bester
Typeset in 10 on 12 pt Times Roman by Studiographix, Cape Town
Colour separations by Hirt & Carter, Cape Town
Printed and bound by National Book Printers, Goodwood

ISBN 0 7981 2372 9

# CONTENTS



*An asterisk (*) after the name of a dish indicates that the recipe appears elsewhere in the book. Consult the index to find out on which page.*

# INTRODUCTION

Poultry, and more specifically chicken, is one of the most versatile and economical food sources available to us. Nearly every form of preparation and cooking can be used, and poultry takes very well to sauces, herbs and wines. The recipes in this book amply demonstrate just how adaptable poultry is, and how acceptable, not only for family meals, but also for special occasions.

In this introduction the most important techniques for preparing poultry, as well as other useful information for selecting and cooking poultry are given.

## Types of poultry

*Chickens* are the most freely available. They can be bought fresh, chilled or frozen; packed in marinades such as lemon and honey, curry or citrus; whole, in portions or as convenience foods. Portions are generally more expensive per kilogram than whole chickens. Free-range chickens, available from some supermarkets, have the best flavour and are the most tender, but frozen chickens are every bit as nutritious. Chickens weigh 1,5 to 2,5 kg, and will feed 4 to 8 people.

*Cornish hens* are larger, tougher birds which require longer cooking and are best pot roasted. They are sometimes available from supermarkets and butchers. They weigh 2 to 3 kg and will feed 6 to 8 people.

*Boiling fowls* are older birds which require longer cooking to ensure tenderness. They are used primarily for boiling or making stock. They weigh 2 to 3 kg and will feed 6 to 8 people.

*Poussins (baby chickens)* are becoming more easily obtainable. The flesh is very tender, and they are best roasted whole, or prepared whole in casseroles or stews. Poussins weigh 500 to 750 g and one should be allowed per person.

*Ducks* are becoming more popular, and are generally available whole (frozen) in supermarkets. They are inclined to be fatty, and care should be taken in the preparation method chosen. Ducks weigh 1,5 to 3,5 kg, and will feed 4 to 8 people.

*Geese* can sometimes be found in supermarkets, but generally only over the festive season. People who have access to farm-reared geese, may consider them worth trying as an alternative for turkey. Younger birds are more tender than the older ones, which tend to be greasy, but geese always require longer cooking times. They weigh 4,5 to 8 kg and will feed 8 to 12 people.

*Turkeys* are the largest of the domestic birds which are sold, usually frozen, in supermarkets. They are often self-basting, which saves time and effort and produces good results. Although available mainly over the festive season, they are becoming easier to obtain. Many supermarkets now also sell turkey steaks (which are marvellous panfried or grilled) and turkey mince. Turkeys weigh 2 to 14 kg and will feed 8 to 20 people.

*Guinea fowls* are not often seen in supermarkets or other stores, and are generally only available to people who have contacts on a game farm, though some butchers specialise in game birds. The flesh, like that of all game birds, is lean and it is necessary to add some fat (in the form of pork fat or bacon) to help keep the meat moist during cooking. They weigh 1 to 1,5 kg and will feed about 4 people.

*Pheasants* are obtainable mainly from a game farm or butcher specialising in game birds. The same comments as for guinea fowls apply.

*Muscovy ducks* are also difficult to obtain, and are best roasted (with added fat) or in a casserole. They range in size from 1 to 1,5 kg and will feed 4 to 6 people, depending on the dish.

*Pigeons* are not freely available either, but very tasty. They can weigh up to 500 g and one bird per person should be allowed.

*Quail.* The same comments as for pigeons apply.

## The nutritional value of poultry

Poultry is an excellent source of high-quality protein. The light meat, such as the breast, is more tender than the dark meat, and also has a lower fat content. The fat is deposited mainly in thick layers under the skin and in the abdominal cavity, which makes it easy to remove for those following a low-cholesterol diet. The fat content of goose, duck and turkey is higher than that of chicken, and the fat has a distinct flavour, which is stronger in older birds. Cooked simply, poultry is easily digestible, which makes it an ideal food for young children and older people.

## Selecting poultry

Consider these points when choosing poultry:

*Chicken.* The breast should be broad and have plump, firm white flesh. The end of the breastbone should be soft and flexible – if it is firm, the bird is older and will not be as tender. Look for white or cream unbroken skin – if the skin is thick and coarse, it is an older bird. A blue discoloration indicates deterioration in quality. Boiling fowls have a slightly yellow tinge to the skin.

*Turkey.* The breast should be plump and well formed and the skin white to slightly bluish and unbroken. The legs should be small in proportion to the size of the bird and the turkey should have a clean, fresh smell.

*Duck and goose.* The young bird has a well-formed breast; if not, it is an older bird. Ducks should have soft white flesh and a fresh smell. The skin should be unbroken and there should be no sign of stickiness.

*Frozen poultry.* When buying prepacked frozen poultry, avoid birds that are limp or contain pinkish ice – signs that they had defrosted and been re-frozen.

## Storing poultry

All poultry should be treated as perishable, whether bought fresh, chilled or frozen. Prepacked fresh and chilled birds should be unwrapped and stored, lightly covered, in the refrigerator for not longer than 3 days. If they are to be stored for longer, wrap in plastic wrap, overwrap and freeze. Frozen poultry should be stored in the freezer until required, overwrapped in plastic wrap.

Chicken, turkey and guinea fowl (uncooked and unstuffed) may be stored in the freezer for up to 6 months and goose and duck for 2 to 3 months. Stuffed birds may be stored in the freezer for up to 2 months.

After freezing, thaw for at least 24 hours in the refrigerator or at room temperature for 7 to 8 hours before use. Frozen poultry can also be thawed in the microwave oven – follow the manufacturer's instructions.

## Boning poultry

Boning may look difficult, but it is not really complicated. The method is the same for all poultry.

*Removing the wishbone*
Pull the flap of skin from the bird's neck down around the shoulders, turning it inside out until you can locate the wishbone with your fingers. Slit just deep enough into the surrounding flesh with a small, sharp knife to expose the wishbone fully. Snap the wishbone from the shoulder joints (1).

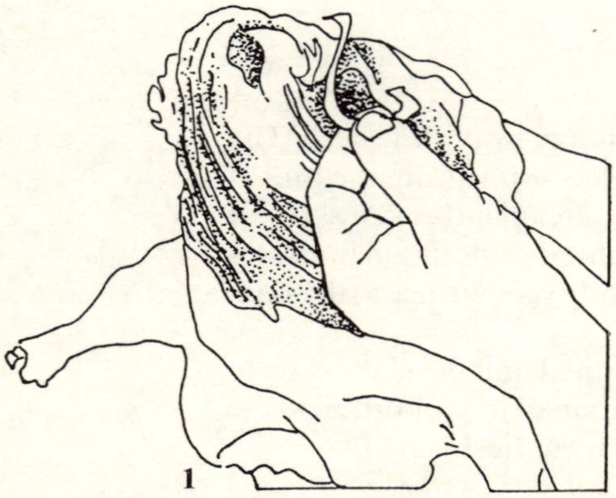

1

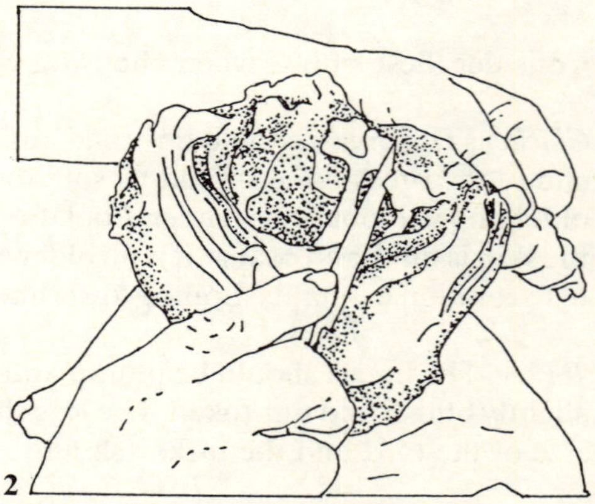

2

*Freeing the wings*
Pull back one wing until you have exposed the sinew that holds it to the collarbone and shoulder blade. Cut through the sinews, and either pull out the bone or leave it intact (2). Repeat for the other wing.

*Snapping the collarbones from the breastbone*

One end of each collarbone is attached to the corresponding shoulder blade and the other end is joined to the breastbone by a weak seam of cartilage. Clear the surrounding flesh from the collarbone with your fingers, scraping with the knife where necessary, and free them from the breastbone (3).

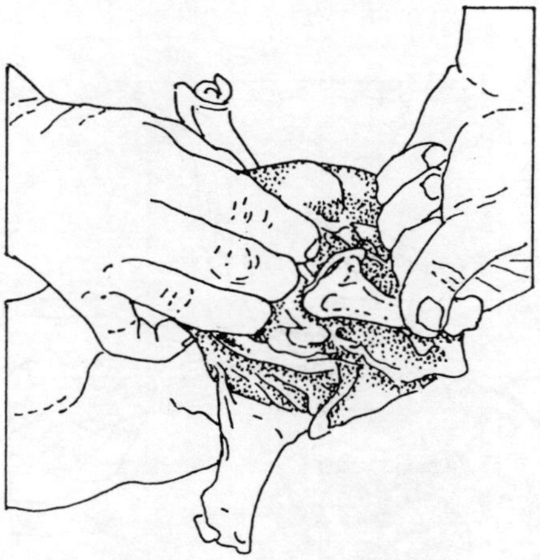

3

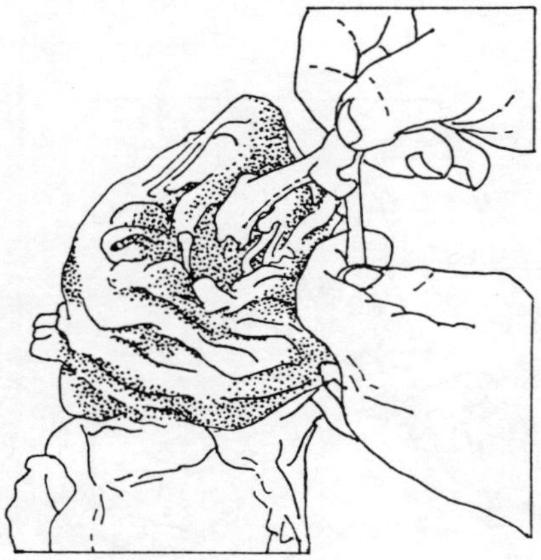

4

*Removing the collarbones from the shoulder blades*

Each collarbone is now attached to a shoulder blade by a thin connecting joint. Pare back the flesh around one joint to expose it fully, then remove both bones by pulling steadily on the joint (4). If the bone breaks, cut out any remaining pieces with the knife. Repeat for the other collarbone and shoulder blade.

*Exposing the carcass*

Pull back the flesh around the shoulders to reveal the top of the carcass – rib cage, breastbone and backbone. Working towards the legs, scrape the flesh from the carcass using your fingers and the knife (5). To avoid damaging the skin, always cut towards the carcass. Where the backbone is attached by the cartilaginous tips of the vertebrae, slice through the tips but leave them in the flesh.

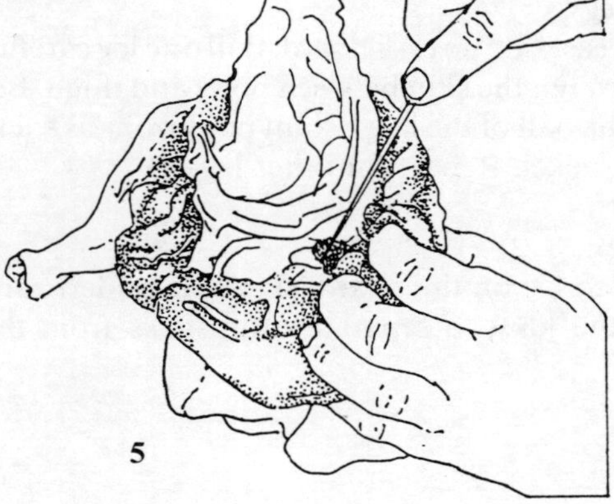

5

*Freeing the breastbone*
When the knife reaches the legs, pop the bones from the ball-and-socket joints where they join the spine and cut through the connecting cartilage. Either leave the leg bones in place or peel back the flesh slightly and pull them out. Continue peeling back the flesh on the carcass until you reach the end of the breastbone – a thin strip of cartilage connects it to the body. Cut through this to free the breastbone (6).

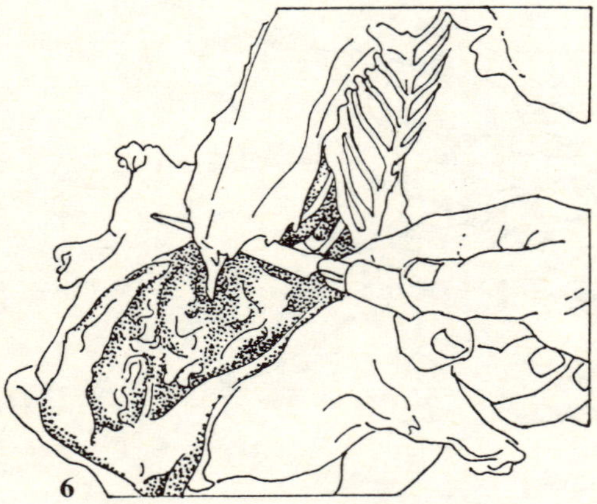
6

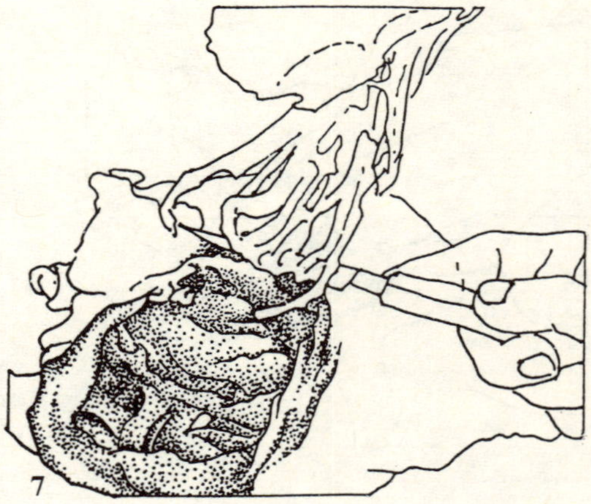
7

*Removing the carcass*
The flesh will now be almost completely peeled away from the carcass. Lift up the carcass and cut through the backbone at the tail, leaving behind the last 3 to 4 vertebrae (7). If the lower ribs still remain in the flesh, cut them out. The bird is now ready for stuffing.

## Jointing poultry

This method of jointing a chicken, duck or game bird approximately the same size, yields 10 portions and there is no wastage. Strong kitchen scissors may be used instead of a knife.

*Removing the legs*
Place the bird breast side up on a board. Pull one leg carefully away from the body and cut through the skin between body and thigh. Bend the leg firmly outwards until the ball of the thigh joint pops from its socket. Cut down between ball and socket. Repeat for other leg.

*Dividing the legs*
Place one chicken leg on the board, skin side underneath, and cut firmly down through the joint to separate drumsticks from thighs. Repeat for other leg.

*Removing the wings*
Press one wing against the body so that both parts of the joint are visible beneath the skin. Cut between the ball and socket of the joint, then pull the wing outwards and cut through the skin at the base of the wing. Repeat for the other wing.

*Splitting the carcass*
Place the knife blade inside the cavity and pierce one side between shoulder joint and rib cage. Cut towards you, parallel to the backbone, and carefully slit the rib cage. Repeat on the other side of the bird.

*Separating the breast*
Pull the breast away from the back to expose the shoulder blades. Cut down between them to detach the breast section. Divide the back by cutting, in the width, across the spine where the rib cage ends.

*Halving the breast*
Place the breast skin side up on the board. Using a strong, steady pressure on the knife, cut through the breastbone on one or the other side of its keel. Larger birds, such as turkeys or geese, can be cut again in the width to yield 4 to 6 breast portions.

## Trussing poultry

Poultry is usually trussed for roasting to keep the legs and wings close to the body and prevent their drying out and becoming tough.

After stuffing the bird, tie the ends of the drumsticks and the tail end together with string. Lift and twist the wings so that they lie flat across the back.

## Roasting times

| Kind | kg | Stuffed | Unstuffed | Oven temperature |
|---|---|---|---|---|
| Duck, goose and chicken | 1-2 | 80 min./kg | 60-80 min./kg | 190 °C for 30 min. then 160 °C |
| | 2,5-3 | 80 min./kg | 50-60 min./kg | |
| Turkey | 3-5 | 60 min./kg | 50-60 min./kg | 190 °C for 30 min. then 160 °C |
| | 5,5-6 | 60 min./kg | 50 min./kg | |
| | 6,5 up | 60 min./kg | 40 min./kg | |

*Notes:*
Use these roasting times unless specified otherwise in recipes.
Game birds are tougher and require longer cooking. Follow instructions in recipe.

## Oven temperatures

| Celsius °C | Fahrenheit °F |
|---|---|
| 100 | 200 |
| 120 | 250 |
| 140 | 275 |
| 150 | 300 |
| 160 | 325 |
| 180 | 350 |
| 190 | 375 |
| 200 | 400 |
| 220 | 425 |
| 230 | 450 |
| 240 | 475 |
| 260 | 500 |
| 280 | 550 |

## Metric volume equivalents

| | | |
|---|---|---|
| ¼ teaspoon | = | 1 ml |
| ½ teaspoon | = | 2 ml |
| 1 teaspoon | = | 5 ml |
| 1½ teaspoons | = | 7 ml |
| 2 teaspoons | = | 10 ml |
| 1 tablespoon | = | 15 ml |
| 2 tablespoons | = | 30 ml |
| 3 tablespoons | = | 45 ml |
| 4 tablespoons | = | 60 ml |
| ¼ cup | = | 65 ml |
| ½ cup | = | 125 ml |
| ¾ cup | = | 200 ml |
| 1 cup | = | 250 ml |
| 1½ cups | = | 375 ml |
| 2 cups | = | 500 ml |
| 3 cups | = | 750 ml |

## Using wine with poultry

Nearly all kinds of wine go well with poultry and game birds. Red wine, in particular, goes well with game birds of all descriptions and with chickens, in casseroles, stews, fried and baked dishes. Fortified wines such as port and sherry add a warm flavour to marinades; white and red wines, including sparkling wines, are also used.

Dry wines are generally used in savoury dishes, though a sweeter wine such as hanepoot or muscadel can impart an unexpected note. Soups are also enhanced by the addition of a dry or medium white or red wine, or sherry, while pâtés are greatly improved if sherry is added.

Traditionally, white wine is served with a poultry dish, but with hearty casseroles or roasts a full-bodied red is perfectly acceptable.

## Using herbs and spices with poultry

Nearly all herbs and spices can be used to enhance poultry dishes. The most often used ones are:

*Basil.* An aromatic, sweet herb, generally used if tomatoes are included.

*Bay leaves.* Very strongly flavoured leaves, used in marinades and bouquets garnis.

*Chilli.* Small pungent fruit pods with a very strong flavour, used fresh or dried, or ground into cayenne pepper. They are used in curries and other spicy dishes.

*Chives.* Plants resembling young spring onions, with a similar flavour. They are used as garnishes or in soups, casseroles, stews and dips.

*Cloves.* The dried flower buds of the clove tree, used in marinades, pickling mixtures, stews and casseroles.

*Coriander.* The seeds are used in curry powder and the leaves as a garnish for soups and salads.

*Fennel.* Looks like a bulbous celery plant and has a strong aniseed flavour. Both the tips and seeds are used as flavourings in stews and casseroles or as garnishes.

*Garlic.* Bulbous plant whose individual pungent cloves are used to flavour a wide variety of dishes.

*Ginger.* The root of a perennial plant which is used fresh or dried and ground in sweet and savoury dishes, especially stir-fries. The young roots may also be preserved in syrup or candied.

*Marjoram.* An aromatic, sweet herb used in soups, stews and stuffings.

*Mint.* A plant with strongly aromatic leaves, which have a variety of uses in savoury dishes.

*Mustard.* The seed of the mustard plant, dried and ground to a powder. Used as a condiment.

*Nutmeg.* The aromatic kernel of the fruit of the nutmeg tree, usually ground, which adds a strong, spicy flavour to marinades, stews and casseroles.

*Oregano.* An aromatic herb used in soups, stews and salads, particularly when tomatoes are included.

*Paprika.* Ground, dried, red sweet peppers, used in dishes with a very spicy flavour.

*Parsley.* A tart herb with a wide variety of uses, in soups, stews, casseroles, sauces and as a garnish.

*Rosemary.* A strongly aromatic herb that enhances stuffings and sauces.

*Saffron.* Imparts flavour and a yellow colour to rice and stuffings.

*Sage.* Silver-white leaves with a sweet aroma, used mainly in stuffings.

*Tarragon.* A sweet-smelling herb that goes particulary well with roast chicken and in stir-fried dishes.

*Thyme.* A small-leafed aromatic herb used primarily in stews and casseroles.

*Turmeric.* Part of the pepper plant which is dried and ground for use in curries and other dishes to impart a yellow colour.

# STARTERS

Poultry makes a delightfully light start to the meal, and because it is so versatile, the range of dishes is wide indeed. We've put together a selection of dishes that will give a choice of old favourites, such as stuffed avocado which gets a tangy new flavour with a savoury chicken filling, and decidedly new ideas, such as the Smoked breast of chicken and avocado terrine. Both show just how deliciously different these same main ingredients can be, prepared in a way that suits today's taste for the fresh and light. Phyllo twists with chicken, pine kernels and feta cheese demonstrate this trend perfectly – light-as-air pastry with a delicate blend of flavours in the filling.

## CHICKEN GOUGÈRE

Serves 8

**Pastry**

*140 g cake flour*
*250 ml water*
*100 g chilled butter, diced*
*5 ml salt*
*4 eggs*
*125 g Gruyère cheese, grated*

**Chicken and artichoke filling**

*2 carrots, cut into julienne strips*
*2 leeks, sliced*
*15 ml olive oil*
*3 cooked artichoke hearts, sliced*
*30 ml chopped fresh parsley*
*1 kg cooked chicken, chopped*
*15 ml cake flour*
*75 ml softened butter*
*200 g smooth cottage cheese or ricotta cheese*
*salt and freshly ground white pepper*

Sift the flour onto greaseproof paper.

Bring the water to the boil, add the butter and salt. Remove the saucepan from the stove as soon as the butter has melted. Add the flour all at once and mix well.

Return the saucepan to the stove and beat with a wooden spoon until the paste no longer sticks to the sides of the pan.

Remove from the stove and add the eggs one by one, beating vigorously after each addition, until the paste is smooth. Stir in 100 g of the cheese.

Butter a baking sheet and drop teaspoonfuls of the paste onto it to make a 23 cm diameter ring. Sprinkle the remaining cheese on top and bake at 200 °C for 1 hour, or until well risen.

Meanwhile, make the filling. Sauté the carrots and leeks in the oil until softened – 5-7 minutes. Stir in the artichokes and parsley and cook for 1-2 minutes. Add the chicken and cook until heated through – about 5 minutes.

Mix the flour and butter to a paste and add gradually to the pan, stirring until smooth and thickened. Stir in the cottage or ricotta cheese, season to taste and cook until heated through – about 10 minutes.

Place the gougère on a serving platter and pour the filling into the centre of the ring. Serve with a green salad and avocado vinaigrette dressing.

## SPICY CHICKEN PASTRIES

Serves 6

*500 g flaky pastry*
*1 egg, beaten*

**Fruity chicken filling**
*180 g finely minced cooked chicken*
*15 ml finely chopped seedless raisins*
*30 ml fruit chutney*
*5 ml Worcester sauce*
*salt and freshly ground black pepper*

**Garnish**
*fennel tips*

First make the filling. Combine the chicken and raisins with the chutney and Worcester sauce and season to taste with salt and pepper.

Divide the pastry into 6 equal portions and roll each out into a small square. Place a sixth of the chicken filling in the centre of each square.

Wet the edges of the pastry and fold over to make triangles. Pinch the edges together to seal them.

Brush the pastry triangles with beaten egg and place them on a baking sheet. Bake at 230 °C for 10-15 minutes, or until golden and puffed up. Garnish with fennel tips and serve.

## CHICKEN WINGS IN PERI-PERI MARINADE

Serves 6

*30 ml butter, margarine or sunflower oil*
*6-12 chicken wings*
*10 ml cornflour*

**Peri-peri marinade**
*1 medium onion, grated*
*1 small chilli, finely chopped or 5 ml peri-peri powder*
*50 ml sunflower oil*
*15 ml fresh lemon juice*
*1 clove garlic, crushed*
*2 ml paprika*
*5 ml salt*

**Garnish**
*lemon wedges*

First make the marinade. Combine the marinade ingredients and marinate the chicken wings for 2-4 hours. Remove from the marinade.

Heat the butter, margarine or oil in a frying pan and fry the chicken wings, turning often, until golden and crisp – about 10 minutes.

Thicken the marinade with the cornflour and bring to the boil. Continue cooking, stirring occasionally, until thick and shiny.

Serve the marinade with the chicken wings, garnished with lemon wedges and accompanied by buttered toasted French bread rounds.

*Note:*
Instead of being fried, the chicken wings may be braaied over the coals or grilled in the oven.

## SPRING ROLLS WITH CHICKEN STUFFING

Serves 6

*12 sheets frozen spring roll pastry, thawed*
*sunflower oil*

**Chicken stuffing**
*500 g minced chicken, cooked*
*100 ml sliced mushrooms*
*30 g rice vermicelli*
*2 cloves garlic, crushed*
*15 ml chopped onion*
*2 eggs, lightly beaten*
*250 ml bean sprouts*
*5 ml salt*

First make the stuffing. Combine all the ingredients for the stuffing well.

Divide the stuffing between the pastry sheets. Roll up the sheets, tucking in the edges as you roll so that the filling is completely enclosed.

Deep-fry in hot oil until crisp and lightly golden, 10-15 minutes. Do not allow to brown too much.

Serve with white rice and a salad of shredded Chinese cabbage leaves and fresh or canned litchis; or sliced, as a starter, with soy sauce as a dip.

*Note:*
If spring roll pastry is unavailable, use small thin crêpes instead.

## CHOUX PUFFS WITH CHICKEN AND BACON FILLING

Makes 24

**Pastry**
*250 ml water*
*90 g butter*
*5 ml salt*
*250 ml cake flour*
*4 large eggs*

**Chicken and bacon filling**
*125 g rindless bacon, finely chopped*
*100 g oyster or button mushrooms, sliced*
*1 small onion, chopped*
*30 ml butter*
*15 ml cake flour*
*250 ml cream*
*15 ml chopped fresh rosemary*
*125 g cooked chicken, chopped*
*100 ml chopped green pepper*

First make the pastry. Place the water, butter and salt in a saucepan and bring slowly to the boil, making sure that the butter melts before the mixture boils.

Add the flour all at once and stir very well with a wooden spoon until the mixture comes away from the sides of the saucepan.

Remove from the stove and cool slightly before beating in the eggs, one at a time.

Spoon scoops of pastry onto a greased baking sheet and bake at 200 °C for 15-20 minutes. Reduce the temperature to 180 °C and bake for a further 15 minutes. Remove and cool on a wire rack.

To make the filling, sauté the bacon, mushrooms and onion in butter until the onion and mushrooms are softened and the bacon is just crisp – 5-10 minutes.

Stir in the flour and cook for 1 minute, while stirring.

Add the cream and rosemary and cook, while stirring, until the sauce thickens.

Add the chicken and green pepper and heat through.

Split the choux puffs and fill them with the chicken filling.

Serve immediately, garnished with a sprig of rosemary.

*Note:*
Fill as many puffs as are needed and freeze the remainder for later use. To serve after freezing, heat the thawed puffs in the oven at 180 °C for 5 minutes, fill and serve.

# PHYLLO TWISTS WITH CHICKEN, PINE KERNELS AND FETA CHEESE

Makes 16

*4 sheets phyllo pastry, each 40 x 30 cm*
*75 ml clarified butter*
*30 ml fine fresh bread crumbs*

**Chicken, pine kernel and feta cheese filling**
*10 ml sunflower oil or butter*
*4 chicken breast fillets, cubed*
*125 g feta cheese, crumbled*
*50 ml pine kernels*
*5 ml salt*
*15 ml snipped chives*

First make the filling. Heat the oil or butter in a frying pan and stir-fry the chicken in it for 2-3 minutes.

Remove the chicken with a slotted spoon and combine it with the feta cheese, pine kernels, salt and snipped chives.

Set aside, covered, to cool.

Put 1 sheet of phyllo pastry on a work surface, wipe it lightly with clarified butter and sprinkle it with 15 ml bread crumbs. Lay another sheet of pastry on top and brush it lightly with some of the remaining clarified butter.

Cut the sheets into 8 squares and spoon about 10 ml filling onto each. Gather the corners of the pastry together over the filling and twist them gently to seal.

Place the twists on a lightly greased baking sheet.

Repeat for the remaining pastry and filling.

Bake the twists at 200 °C for 10-12 minutes, and serve warm.

*Notes:*

To clarify butter, heat unsalted butter in a saucepan over low heat until clear, skimming foam from the surface occasionally. Pour through a very fine sieve lined with thin muslin or cheesecloth and discard the sediment. Use the clear liquid as required. (It will keep in the refrigerator for up to 6 weeks.)

Ready-made phyllo pastry may be bought from supermarkets and some delicatessens. If you prefer, 250 g puff pastry may be used instead. Omit the clarified butter and bread crumbs and make the twists without layering the pastry.

## FRESH PEA AND CHICKEN TIMBALES

Serves 6

*350 g fresh peas*
*500 g finely minced cooked chicken*
*4 eggs*
*250 ml sour cream*
*65 ml grated Parmesan cheese*
*50 g smooth cottage cheese*

**Tomato coulis**

*15 ml sunflower oil*
*1 clove garlic, crushed*
*1 onion, chopped*
*1 x 410 g can tomatoes*
*2 ml sugar*

**Garnish**

*finely shredded spring onions*

Steam the peas until just tender, about 5 minutes, and drain. Purée until smooth in a blender or food processor, then push through a fine sieve.

Stir in the chicken and blend well.

Combine the eggs, sour cream and cheeses in a bowl and whisk until well combined.

Stir in the pea purée and pour the mixture into 6 lightly greased 125 ml capacity moulds. Cover each with a piece of lightly greased foil.

Place moulds in a baking dish with enough hot water to reach halfway up the sides of the moulds. Bake at 180 °C for 30 minutes, or until set.

Meanwhile, make the tomato coulis. Heat the oil in a saucepan, add the garlic and onion and cook until softened – about 5 minutes. Add undrained crushed tomatoes and sugar and cook, stirring, for 5 minutes.

Purée in a blender or food processor until smooth, then push through a fine sieve.

Remove the moulds from the baking dish and allow to stand for a few minutes before turning the timbales out onto serving plates.

Serve with tomato coulis poured round the timbales, garnished with shredded spring onions.

## CURRIED DUCK LIVERS

Serves 4

*450 g duck livers*
*60 ml butter*
*5 ml cake flour*
*10 ml curry powder*
*150 ml cream*
*10 ml fruit chutney*
*15 ml fresh lemon juice*

Trim the livers and cut them into large pieces.

Heat the butter in a frying pan and stir-fry the livers until cooked but still pink inside – about 5 minutes. Remove the livers and keep them warm.

Blend the flour and curry powder with a little of the cream and when smooth, whisk in the remaining cream and the chutney. Stir into the pan juices and cook while stirring, for 5-7 minutes or until smooth.

Return the livers to the pan, add the lemon juice and heat through. Serve on buttered whole-wheat toast.

## AVOCADO HALVES WITH CHICKEN FILLING

*A marvellously piquant variation on the usual seafood-stuffed avocado served as a first course*

Serves 4

*2 large ripe avocados, halved and stoned*
*fresh lemon juice*

**Chicken and litchi filling**
*200 g cooked chicken, cut into slivers*
*5 fresh or canned litchis, stoned and coarsely chopped*
*2 pieces preserved ginger, finely chopped*
*30 ml vinaigrette dressing*
*30 ml chopped raw cashew nuts*
*2 spring onions, thinly sliced*

**Garnish**
*freshly ground black pepper*
*finely chopped fresh tarragon*

Sprinkle the avocado halves with lemon juice and refrigerate until needed.

To make the filling, combine all the ingredients well and use to fill the cavities in the avocado halves.

Sprinkle with pepper and tarragon and serve.

## CHICKEN LIVER PÂTÉ

Makes 750 ml

*200 g butter*
*500 g chicken livers*
*250 g mushrooms, finely chopped*
*1 large green pepper, seeded and finely chopped*
*1 small onion, finely chopped*
*75 ml dry sherry*
*1 clove garlic, crushed*
*4 drops Tabasco sauce*
*5 ml salt*

Melt 65 ml of the butter in a large frying pan and sauté the livers, mushrooms, green pepper and onion for 10 minutes.

Add the sherry, garlic, Tabasco sauce and salt and remove from the stove.

Allow the mixture to cool slightly, then purée it until smooth in a blender or food processor with the remaining butter.

Pack the pâté tightly into an earthenware crock and chill, covered, for at least 3 hours.

Serve with Melba toast or toast fingers.

## SMOKED CHICKEN AND ORANGE SALAD

Serves 4

*2 large oranges, peeled and all pith removed*
*1 large mango, peeled and stoned*
*chicory leaves*
*4 smoked chicken* breasts, skinned and chopped*

**Dressing**
*100 ml mayonnaise*
*125 ml sour cream*
*30 ml orange juice*
*pinch castor sugar*
*salt and freshly ground black pepper to taste*

**Garnish**
*30 ml lightly toasted walnuts or pecan nuts*
*parsley sprigs*

Cut the oranges into neat segments and slice the mango into crescents. Arrange the orange, mango and chicory on 4 plates and top with the chicken.

Garnish the salad with the nuts and parsley.

Combine the dressing ingredients well in a screw-topped jar and serve separately.

*Note:*
Instead of oranges and the mango, serve with 4 ripe figs, quartered, and seedless green grapes.

## CHICKEN MOUSSE

Serves 4-6

| | |
|---|---|
| *1* | *small onion, finely chopped* |
| *15 ml* | *butter or margarine* |
| *30 ml* | *cake flour* |
| *375 ml* | *milk* |
| *15 cm* | *piece English cucumber, grated* |
| *250 g* | *cooked chicken, minced* |
| *120 g* | *cooked ham, minced* |
| *10 ml* | *mint sauce or jelly* |
| | *pinch each salt and freshly ground white pepper* |
| *15 ml* | *gelatine powder* |
| *30 ml* | *water* |
| *150 ml* | *whipped cream* |
| *20 ml* | *dry sherry* |

**Garnish**
*very thinly sliced English cucumber*

Sauté the onion in the butter or margarine in a large saucepan until tender but not brown - about 5 minutes.

Stir in the flour and cook for 1 minute, stirring constantly.

Remove the saucepan from the stove and gradually stir in the milk. Return the saucepan to the stove and bring to the boil, stirring constantly.

Boil for 1 minute, then remove from the stove.

Stir in the grated cucumber, chicken, ham, mint sauce or jelly and seasoning.

In a large bowl, dissolve the gelatine powder in the water over a pan of hot water.

Stir in the chicken mixture and allow to cool slightly.

Stir in the cream and sherry and pour the mixture into a mould sprayed with non-stick cooking spray.

Refrigerate until set.

Turn out onto a platter and serve garnished with cucumber slices and accompanied by lengths of chopped celery.

*Note:*
Leftover turkey or guinea fowl may be used instead of chicken.

## ROAST CHICKEN WITH DIPPING SAUCES

Serves 4-6

*1 kg roast chicken, boned and cut into bite-sized pieces*
*dipping sauces of your choice**

Place the chicken on a serving platter and serve dipping sauces separately.

## CINNAMON YOGHURT DIPPING SAUCE

Makes 400 ml

*375 ml natural yoghurt*
*10 ml brown sugar*
*5 ml ground cinnamon*
*5 ml ground ginger*
*10 ml fresh lemon juice*

Whisk together the yoghurt and sugar until the sugar has dissolved.

Add the cinnamon, ginger and lemon juice and combine well.

*Note:*
Lemon-flavoured yoghurt may be used instead of natural yoghurt. Leave out the sugar and lemon juice.

## CUCUMBER SOUR-CREAM DIPPING SAUCE WITH FRESH MINT

Makes 650 ml

*250 ml sour cream*
*375 ml grated cucumber, drained*
*1 clove garlic, crushed*
*10 ml fresh lemon juice*
*15 ml chopped fresh mint*

Combine all ingredients well. Chill for 1 hour before serving.

## MANGO CHUTNEY AND CORIANDER DIPPING SAUCE

Makes 400 ml

*45 ml mango chutney*
*50 ml chopped pecan nuts*
*10 ml fresh lemon juice*
*250 g homogenised cottage cheese or ricotta cheese*
*50 ml natural yoghurt*
*5 ml curry powder*
*30 ml chopped fresh coriander leaves*

Blend all the ingredients well. Chill for 1 hour before serving.

*Note:*
If fresh coriander leaves are unavailable, use 5 ml ground coriander instead.

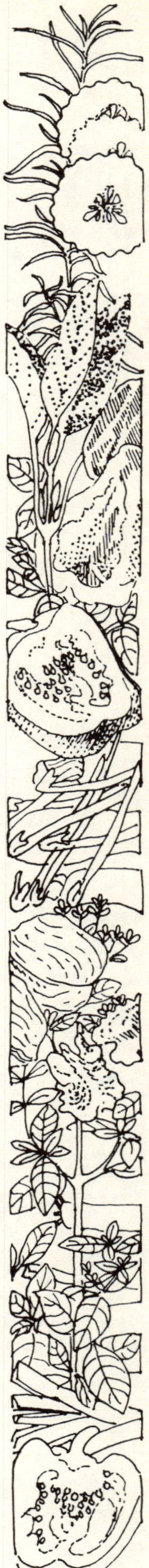

## TANGY TOMATO DIPPING SAUCE

Makes 500 ml

*1 x 410 g can plum tomatoes, drained and chopped*
*1 green pepper, seeded and chopped*
*1 clove garlic, crushed*
*5 ml salt*
*2 ml sugar*
*2 ml coarsely ground black pepper*
*10 ml olive oil*
*1 small onion*
*15 ml chopped fresh basil*
*2 ml Tabasco sauce*
*1 slice day-old white bread, crusts removed*

Blend all ingredients in a blender or food processor to form a thick purée.

Chill well before serving.

## SMOKED BREAST OF CHICKEN AND AVOCADO TERRINE

Serves 8

| | |
|---|---|
| *20 ml* | *gelatine powder* |
| *300 ml* | *chicken stock** |
| *2* | *large ripe avocados, peeled, stoned and mashed* |
| | *salt and freshly ground black pepper* |
| | *fresh lemon juice* |
| | *Tabasco sauce* |
| *150 ml* | *mayonnaise* |
| *250 ml* | *thick cream* |
| *500 g* | *smoked chicken* breast, thinly sliced* |
| | *butter lettuce leaves* |

**Garnish**

*thin English cucumber slices*

Dissolve the gelatine in half the stock, heat and, when clear, add to the remaining stock. Mix well.

Season the avocados with salt, pepper, lemon juice and Tabasco sauce to taste.

Beat in the mayonnaise, then the gelatine mixture. Whip the cream and fold it into the avocado mixture.

Line a greased 1,5 ℓ loaf pan with plastic wrap and layer a third of the chicken on the bottom. Cover with half the avocado mixture, another layer of chicken, the remaining avocado mixture and the remaining chicken.

Freeze until firm, then unmould, double-wrap the terrine in plastic wrap and return to the freezer until needed.

Thaw the terrine overnight in the refrigerator, unwrap and place on a bed of lettuce leaves on a serving platter. Garnish with cucumber slices.

Serve sliced, with Melba toast and a green salad.

*Note:*
Smoked turkey or fresh chicken poached in stock may be used instead of smoked chicken.

# SOUPS

If you've always thought that Chicken noodle soup is the only poultry soup around – think again. The poultry soups given here range from cold soups such as Chicken vichyssoise and Chilled chicken avocado soup to ribsticking meals in one dish such as our version of Chicken gumbo and Pot au feu. The traditional is catered for, as well as the more unusual and the recipe for Chicken stock is given simply because you'll use it again and again in all kinds of recipes.

## CHICKEN STOCK

Makes 2 ℓ

*1 x 2 kg chicken*
*1 bouquet garni (parsley, thyme, bay leaf)*
*1 clove garlic, crushed*
*2 cloves*
*6 peppercorns*
*10 ml salt*
*3 ℓ water*
*4 carrots, sliced*
*2 leeks, sliced*
*2 stalks soup celery, chopped*

Place the chicken, herbs, spices, salt and water in a large saucepan and bring slowly to the boil.

Simmer, covered, for 1-1½ hours, skimming the scum from the surface when necessary.

Add the vegetables and continue simmering until the chicken is tender, 1½-2 hours.

Skim the fat from the surface. Strain the stock through a fine sieve, discarding whatever will not go through the sieve.

Cool and store in the refrigerator, or freeze until needed.

## CHICKEN NOODLE SOUP

Serves 4-6

*1,5 ℓ chicken stock**
*2 carrots, diced*
*2 stalks celery, finely chopped*
*3 spring onions, finely chopped*
*5 ml salt*
*2 ml freshly ground white pepper*
*150 g mushrooms, thinly sliced*
*50 g uncooked egg noodles*
*150 g cooked chicken, diced*

Bring the stock to the boil in a large saucepan.

Add the carrots, celery, spring onions, salt and pepper and simmer, covered, for 15 minutes.

Stir in the mushrooms and noodles, using a wooden spoon, then add the chicken and simmer for a further 10 minutes.

Serve immediately.

# CHICKEN CREAM SOUP

Serves 4-5

**Béchamel sauce**

*75 ml butter*
*75 g cake flour*
*1 ℓ milk or 500 ml each milk and chicken stock**
*1 small onion, very finely chopped*
*2 ml freshly ground white pepper*
*2 ml grated nutmeg*
*2 ml dried thyme*
*15 ml salt*

**Soup**

*200 ml milk or 250 ml chicken stock**
*500 g skinned uncooked chicken meat*
*100 ml thick cream*

**Garnish**

*fresh parsley or thyme sprigs*

First make the Béchamel sauce. Melt the butter in a large heavy-based saucepan. Add the flour and stir with a wire whisk until it has all been absorbed.

Whisk in the milk or milk and chicken stock and bring to boil over high heat, stirring often.

Sauté the onion in a little butter for 5 minutes, without browning. Add to the sauce with the pepper, nutmeg, thyme and salt.

Simmer the sauce, uncovered, over the lowest possible heat for 1 hour. Do not stir.

Strain through a fine sieve and set aside, discarding whatever does not go through the sieve.

To make the soup, pour the sauce into a saucepan, add the milk or stock and the chicken meat and simmer, uncovered, until the chicken is tender – about 20 minutes. Do not stir.

Purée the soup in a blender or food processor, first on low speed and then on high, until smooth.

Strain through a fine sieve, discarding whatever does not go through. Stir in the cream.

Serve immediately, garnished with the parsley or thyme.

## CHICKEN GUMBO

*Our version of the famous Creole dish, using baby marrows instead of okra*

Serves 8

| | |
|---|---|
| *75 ml* | *butter or margarine* |
| *500 g* | *baby marrows, trimmed and sliced* |
| *1* | *large onion, chopped* |
| *1* | *green pepper, seeded and chopped* |
| *1* | *clove garlic, crushed* |
| *30 ml* | *cake flour* |
| *1,5 ℓ* | *chicken stock** |
| *4* | *large ripe tomatoes, skinned, seeded and chopped* |
| *2* | *sprigs parsley and 1 bay leaf, tied together* |
| *2 ml* | *dried thyme* |
| | *salt and freshly ground white pepper to taste* |
| *3-4* | *skinned chicken breast fillets* |
| *30 ml* | *Worcester sauce* |
| *100 ml* | *thick cream* |
| *100 g* | *cooked rice* |

**Garnish**

*chopped fresh parsley or dill*

Melt 50 ml of the butter or margarine in a frying pan and sauté the baby marrow slices for 2-3 minutes on either side, or until lightly browned. Set aside.

Melt the remaining butter or margarine in a large saucepan and sauté the onion and green pepper for 5 minutes.

Add the garlic and sauté for 1 minute.

Sprinkle the flour over the vegetables and stir until completely absorbed.

Add the stock, stirring vigorously.

Add the baby marrow slices, tomatoes, parsley and bay leaf, thyme, salt and pepper and the chicken breasts.

Bring the soup to the boil, then reduce the temperature and simmer, partially covered, for about 30 minutes or until the meat is tender.

Remove the chicken and, when it has cooled sufficiently to handle, dice the meat and return it to the saucepan.

Add the Worcester sauce and adjust the seasoning if necessary, then stir in the cream.

Place 15 ml cooked rice in each soup plate and ladle the soup over it. Serve immediately, garnished with fresh parsley or dill.

*Right: Smoked breast of chicken and avocado terrine (p. 28)*

# MULLIGATAWNY

Serves 8

| | |
|---|---|
| 2 | *chicken breasts, on the bone* |
| 3 ℓ | *chicken stock** |
| 1 | *large onion, finely chopped* |
| 2 | *medium carrots, finely chopped* |
| 2 | *stalks soup celery, finely chopped* |
| 40-50 ml | *butter or margarine* |
| 1 | *bay leaf* |
| 45 ml | *cake flour* |
| 15-30 ml | *curry powder* |
| 100 g | *cooked long-grain rice* |
| | *salt and freshly ground black pepper* |

**Garnish**

*chopped fresh parsley or coriander leaves*

Place the chicken breasts and the stock in a large saucepan and bring to the boil, then simmer, partially covered, over very low heat until the chicken is tender – about 20 minutes.

Remove the chicken breasts and when cool enough to handle, skin and bone them. Shred the meat and set aside.

Sauté the onion, carrots and celery in the butter or margarine for 5 minutes, then add the bay leaf.

Sprinkle the vegetables with the flour and curry powder and stir very well with a wooden spoon to coat all the vegetables.

Stir the vegetable mixture into the liquid in the saucepan, using a wire whisk to incorporate the flour and curry powder completely.

Bring to the boil, then reduce the temperature and simmer, partially covered, for 30 minutes.

Add the chicken and cooked rice and simmer for 1 minute longer.

Season to taste with salt and pepper.

Serve garnished with fresh parsley or coriander leaves.

*Left: Chicken bouillabaisse (p. 39)*

## CHICKEN VEGETABLE SOUP

Serves 6

*30 ml butter or margarine*
*1 large onion, chopped*
*4 carrots, diced*
*250 ml fresh green peas*
*1 small head cauliflower, broken into florets*
*2 new potatoes, diced*
*250 g green beans, thinly sliced*
*4 radishes, halved*
*1 ℓ chicken stock**
*100 g spinach, finely chopped*
*250 g cooked chicken, diced*
*10 ml salt*
*2 ml freshly ground white pepper*

**Garnish**
*250 ml croûtons**

Heat the butter or margarine in a large saucepan and sauté the onion until golden - about 5 minutes.

Add the carrots, peas, cauliflower, potatoes, beans and radishes, cover with stock and bring to the boil.

Cook until the vegetables are just tender - about 15 minutes.

Add the spinach and cook for another 5 minutes.

Add the chicken and simmer, covered, for 3-5 minutes.

Season and serve immediately, garnished with croûtons.

## AVGOLEMONO SOUP

*Egg and lemon soup*

Serves 6

*2,25 ℓ chicken stock**
*125 ml uncooked rice*
*5 ml salt*
*2 ml freshly ground white pepper*
*2 eggs*
*juice of 1 lemon (fresh)*

**Garnish**
*strips of lemon rind*

Heat the chicken stock to boiling point in a large saucepan.

Add the rice and simmer, covered, for 15-20 minutes, or until the rice is soft. Season with salt and pepper.

Beat the eggs in a small dish and add the lemon juice. Gradually add a little of the hot soup to the egg mixture, beating constantly, then stir the mixture into the soup.

Garnish with strips of lemon rind and serve.

## COCK-A-LEEKIE

Serves 4

| | |
|---|---|
| *750 g* | *leeks, sliced 5 mm thick* |
| | *water* |
| *10 ml* | *salt* |
| *1 x 1,5 kg* | *chicken* |
| *1 ℓ* | *chicken stock** |
| *3* | *sprigs parsley and 1 bay leaf, tied together* |
| *2 ml* | *dried thyme* |
| *50 g* | *pearl barley* |
| *200 ml* | *thick cream, or to taste* |
| | *salt and freshly ground white pepper to taste* |

**Garnish**

*chopped fresh parsley or thyme*

Cover the leeks with cold water and sprinkle with salt. Leave to soak for 10 minutes, then rinse well.

Place the chicken in a large saucepan and add the stock, parsley and bay leaf bundle, thyme and barley.

Simmer, partially covered, for about 40 minutes.

Add the leeks and cook for a further 15 minutes, or until the barley is tender.

Remove the chicken and set it aside. Lift out and discard the parsley bundle. When cool enough to handle, skin and bone the chicken and cut the meat into bite-sized pieces.

Return the meat to the saucepan and heat the soup through. Remove the saucepan from the stove, stir in the cream and season well with salt and pepper.

Reheat to just below boiling point.

Serve immediately, garnished with fresh parsley or thyme.

## OLD-FASHIONED CHICKEN SOUP WITH MEATBALLS AND DUMPLINGS

Serves 6

*1,5 ℓ chicken stock**
*2 leeks, thinly sliced*
*3 carrots, sliced and lightly cooked*
*250 g cooked chicken, cut into slivers*
*salt*

**Meatballs**
*250 g pork, minced*
*salt and freshly ground black pepper*
*15 ml cake flour*
*1 egg*
*100 ml cream*
*500 ml chicken stock**

**Dumplings**
*100 ml water*
*75 ml butter*
*75 ml cake flour*
*5 ml salt*
*2 eggs, lightly beaten*

**Garnish**
*parsley sprigs*

First make the meatballs. Mix all the ingredients except stock well and, with floured hands, shape into small balls. Place a few balls at a time in the barely simmering chicken stock and simmer for 4-8 minutes. Remove the balls with a slotted spoon and dip them quickly into cold water.

Bring the water for the dumplings to the boil with the butter. Sprinkle in the flour and mix until the dough is heated through. Chill the dough briefly in the refrigerator, then mix in the salt and eggs, a little at a time. Shape into small dumplings using two spoons and lower into a saucepan of simmering salted water, a few at a time. Cook until they float to the surface, 5-7 minutes.

Remove the dumplings with a slotted spoon and refresh them in cold water.

Heat the 1,5 ℓ chicken stock until boiling and add the strained stock from the meatballs. Add the leeks and boil for 4-5 minutes, then add the carrot, chicken, meatballs and drained dumplings. Heat through well, but do not allow to boil. Season with salt, if necessary.

Serve immediately, garnished with parsley sprigs and accompanied by buttered whole-wheat toast.

*Note:*
The soup, meatballs, dumplings and vegetables can be frozen separately. To serve, thaw and reheat the ingredients together, but do not allow to boil.

## SMOKED-TURKEY CHOWDER

Serves 4

| | |
|---|---|
| *100 g* | *salt pork, diced* |
| *1* | *small onion, finely chopped* |
| *30 ml* | *diced green pepper* |
| *2* | *medium potatoes, diced* |
| *250 ml* | *diced celery* |
| *250 ml* | *cooked whole-kernel corn* |
| *375 ml* | *boiling water or chicken stock** |
| *2 ml* | *salt* |
| *1 ml* | *freshly ground white pepper* |
| *250 ml* | *diced smoked turkey** |
| *1 x 397 g* | *can evaporated milk* |
| *10 ml* | *cake flour* |
| *1* | *canned pimento, drained and finely chopped (optional)* |

Fry the pork slowly in a large saucepan until crisp, 5-10 minutes. Remove the pork and pour off all but 30 ml of the fat, reserving 10 ml of the poured-off fat.

Add the onion and green pepper to the saucepan and sauté for 2-3 minutes.

Add the potatoes, celery, corn, water or stock, salt and pepper and bring to the boil.

Reduce the temperature and simmer, covered, until the potato is tender – about 15 minutes. Add the turkey.

Blend the milk and the reserved pork fat into the flour, then stir the mixture into the soup.

Bring to the boil, stirring continuously.

Stir in the pimento, if used, and the reserved pork and serve immediately.

*Note:*
Cooked chicken may be used instead of the turkey.

## POT AU FEU

*A classic French peasant soup hearty enough to be a meal in itself*

Serves 4

*1 x 1,5 kg chicken*
*500 g chicken carcasses and necks*
*3 ℓ water*
*2 large carrots, chopped*
*2 leeks, sliced*
*1 medium turnip, chopped*
*1 stalk celery, chopped*
*1 large onion, chopped*
*2 sprigs parsley and 1 large bay leaf, tied together*
*2 ml dried thyme*
*salt and freshly ground black pepper*
*1 cut-up marrowbone, tied in a cheesecloth or muslin bag*

Place the whole chicken, carcasses and necks in a large heavy-based saucepan. Add the cold water and bring to the boil over medium heat. Skim off the scum as it appears, continuing to do so until the surface is clear.

Add the vegetables, parsley and bay leaf, thyme, salt and pepper and simmer, partially covered, over very low heat for 1½ hours. (The liquid should be barely bubbling.)

Add the tied-up marrowbone and set aside. Remove and discard the chicken carcasses and necks and the parsley bundle.

Strain the broth through a fine sieve lined with a damp cloth and discard all the vegetables.

Taste the broth and adjust the seasoning, if necessary, then remove as much grease as possible.

Serve the broth with the marrow from the bone spread on hot toast, and the carved chicken separately, with boiled potatoes.

## TURKEY AND PEA SOUP

Serves 8-10

*1 kg leftover turkey, coarsely chopped*
*1 kg cooked ham, diced*
*2 ham bones*
*2 ℓ chicken stock**
*500 g carrots, sliced*
*1 head celery, chopped*
*4 leeks, sliced*
*500 g small onions*
*1 sprig parsley*
*500 g yellow split peas*
*salt*

Place the turkey, ham and ham bones in a large saucepan and add the stock.

Bring to the boil, then reduce the temperature and simmer, covered, over low heat for about 1 hour, skimming the surface when necessary.

Add the vegetables, except the peas, and simmer for 15-20 minutes.

Rinse the peas in cold water and add to the saucepan. Make sure that they are covered by stock. Simmer, covered, over low heat for 30-45 minutes, then taste and season with salt, if necessary.

Serve with buttered brown bread.

## CHICKEN BOUILLABAISSE

Serves 4

*1 x 1,5 kg chicken, cut into portions*
*5 ml ground saffron threads or turmeric*
*30 ml dry sherry*
*200 ml olive oil*
*5 ml salt*
*2 ml freshly ground white pepper*
*2 onions, chopped*
*5 cloves garlic, crushed*
*6 ripe tomatoes, skinned, seeded and chopped*
*10 sprigs fennel*
*30 ml chopped parsley*
*1,5 ℓ chicken stock**
*4 potatoes, thickly sliced*
*1 clove garlic, crushed*
*10 ml olive oil*
*4 slices French bread*

Marinate the chicken portions for 30 minutes in a mixture of saffron or turmeric, sherry, 100 ml olive oil, salt and pepper.

Heat the remaining olive oil in a saucepan and sauté the onions and garlic in it until transparent – about 5 minutes.

Add the tomatoes, fennel, parsley, chicken portions and marinade. Pour in the chicken stock and boil, covered, for 10 minutes.

Add the potatoes and simmer, covered, for about 30 minutes more, or until the potatoes are soft.

Combine the garlic and olive oil and spread on the French bread slices. Toast under a pre-heated grill.

Serve the bouillabaisse with the French bread.

*Note:*
Traditionally, the broth is served first and the chicken portions and vegetables as a second course with the toasted French bread.

## CHILLED CHICKEN AVOCADO SOUP

Serves 4-6

| | |
|---|---|
| 2 | *ripe avocados* |
| 750 ml | *chicken stock** |
| | *juice of ½ lemon* |
| | *salt and freshly ground white pepper* |
| 150 ml | *very finely chopped cooked chicken* |
| 100 ml | *thick cream or natural yoghurt* |

**Garnish**

*chopped fresh parsley or tarragon*

Peel, stone and coarsely chop the avocados.

Add the chicken stock and liquidise to a smooth purée in a blender or food processor. (If you do not have a blender or food processor, force the avocado flesh through a fine sieve and mix well with the chicken stock.)

Add the lemon juice, salt and pepper to taste and cooked chicken, and refrigerate, covered, until lightly chilled.

Stir in the cream or yoghurt and serve garnished with parsley or tarragon.

*Note:*

Do not serve this soup too cold, as it will lose much of its flavour.

## CHICKEN VICHYSSOISE

Serves 6

| | |
|---|---|
| 2 kg | *large potatoes, coarsely chopped* |
| 4 | *leeks, sliced* |
| 1,5 ℓ | *chicken stock** |
| 375 ml | *diced cooked chicken* |
| 250 ml | *cream* |
| 20 ml | *brandy* |
| | *salt and freshly ground white pepper* |

**Garnish**

*snipped fresh chives*

Combine potatoes, leeks and stock in a large saucepan. Bring to the boil, then reduce the temperature and simmer, covered, for 40 minutes. Add the chicken after 30 minutes.

Purée the soup in a blender or food processor and return it to the saucepan.

Stir in the cream and brandy and heat gently without boiling. Season with salt and pepper, if necessary.

Chill in the refrigerator and serve, topped with chives.

*Note:*
This soup is also delicious served hot.

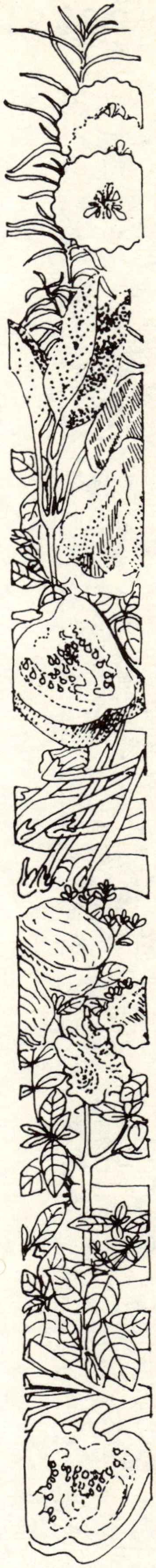

## IDEAS FOR SOUP GARNISHES AND ACCOMPANIMENTS

*Croûtons.* Cube day-old white bread and fry, tossing frequently, in very hot butter or oil for about 1 minute, or until crisp. Very finely chopped fresh herbs of your choice, 1 clove crushed garlic, curry powder, or paprika may be added to the butter before frying the croûtons.

*Melba toast.* Cut white bread very thinly and toast in the oven at 180 °C until crisp and lightly browned.

*Cheese toast.* Quarter slices of bread, spread with butter and top with grated Cheddar cheese. Sprinkle with paprika and grill until the cheese bubbles.

*Vegetable crudités.* Clean and chop vegetables such as carrots, green peppers, cucumber and mushrooms, or cut them into julienne strips; leave cherry tomatoes and cauliflower florets whole.

*Whole-wheat toast fingers.* Cut whole-wheat bread slices into fingers, top with butter mixed with mustard and grill until golden.

# LIGHT MEALS

Whether it's a quick snack or a sandwich, a lunchtime treat or a supper for friends, you'll find a recipe to fit the bill in this selection of quick and easy-to-prepare dishes. The salads are marvellous for midday, and for most of them you can use leftover cooked poultry. Unusual snack ideas, such as Tacos with chicken and cheese filling and Turkey-stuffed pita breads, will keep the compliments rolling in, and you'll find the ideas for sandwich fillings and toppings a great boon when you can't make up your mind what to give the children to take to school. All-time favourites get a new look here, too – green peppers stuffed with chicken and rice, for instance, and cannelloni stuffed with ricotta cheese and chicken.

## DUCK AND ORANGE SALAD WITH SOUR-CREAM DRESSING

Serves 6

*750 g cooked duck, sliced*
*125 ml dry white wine*
*15 ml fresh lemon juice*
*10 ml finely chopped onion*
*5 ml salt*
*5 ml chopped fresh tarragon or chervil*
*15 ml chopped blanched almonds*
*2 stalks celery, trimmed and chopped*
*2 large oranges, peeled with all pith removed*

**Sour-cream dressing**

*50 g cream cheese or smooth cottage cheese*
*125 ml sour cream*
*salt and freshly ground white pepper*
*1 ml castor sugar*
*dash Tabasco sauce*

**Garnish**

*watercress sprigs*

Chop the duck and place it in a bowl with the wine, lemon juice, onion, salt and herbs. Stir well and leave, covered, in the refrigerator for 24 hours, turning from time to time.

Drain off the unabsorbed liquid and stir the almonds and celery into the duck mixture. Seed the oranges and chop the flesh coarsely. Stir into the duck mixture.

To make the sour-cream dressing, mash the cream or cottage cheese with a fork and blend in the sour cream. If it seems too stiff, add a little milk. Season with salt, pepper, sugar and Tabasco sauce.

When ready to serve, pour the dressing over the duck salad and garnish with watercress sprigs.

*Note:*

Cold roast chicken, or smoked chicken or turkey* may be used instead of duck.

## CHICKEN AND APPLE SALAD

Serves 4

*cos or iceberg lettuce leaves*
*300 g cooked chicken, cubed*
*2 hard-boiled eggs, quartered lengthwise*
*3 cherry tomatoes, halved*
*100 g mushrooms, sliced*
*1 Granny Smith apple, cored and cubed*
*10-12 stuffed olives, sliced*
*30 ml fresh lemon juice*

**Herb dressing**
*100 g cream cheese or smooth cottage cheese*
*30 ml cream*
*salt and freshly ground white pepper*
*15 ml chopped fresh chives*
*15 ml chopped fresh parsley*

Line a salad platter with the lettuce leaves. Combine the salad ingredients in a bowl, sprinkle with lemon juice and spoon onto the lettuce leaves.

To make the dressing, mash the cheese with a fork, then whisk in the cream until smooth. Season with salt and pepper, stir in the herbs and pour over the salad.

Serve with buttered beer bread.

*Note:*
Any leftover poultry may be used for this salad.

## DUCK AND PEAR SALAD

Serves 4-6

*cos lettuce leaves*
*2 pears, cored, quartered and sliced lengthwise*
*4-6 cooked duck breasts, skinned*

**Lime and ginger dressing**
*125 ml walnut or sunflower oil*
*1 pear, peeled and cored*
*1,5 cm piece peeled fresh ginger*
*15 ml vinegar*
*15 ml fresh lime or lemon juice*
*15 ml grated lime or lemon rind*
*salt and freshly ground white pepper to taste*

**Garnish**
*125 ml skinned pistachio nuts or pine kernels*
*chopped fresh mixed herbs*

Place the cos lettuce leaves on a serving platter and arrange the pear slices on top.

Slice the duck breasts diagonally and arrange over lettuce and pears.

Purée the dressing ingredients in a blender or food processor until smooth and spoon over duck.

Garnish with nuts and herbs and serve immediately.

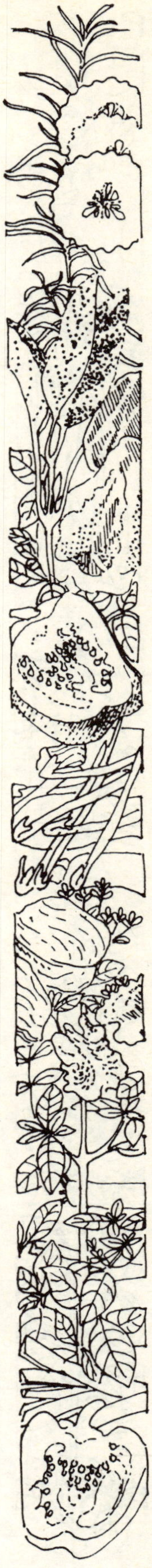

## CURRIED CHICKEN PASTA SALAD

Serves 4-6

*350 g pasta shells*
*butter*
*150 ml mayonnaise*
*30 ml cream*
*10 ml mild curry powder*
*salt and freshly ground white pepper*
*350 g cooked chicken, diced*
*100 g smoked ham or salami, diced*
*2 stalks celery, diced*
*1 small green pepper, seeded and sliced*
*lettuce leaves*

Cook the pasta shells in boiling water until barely tender, about 12 minutes. Drain, rinse and dry, then toss the shells in a little heated butter for 2-3 minutes. Drain.

Combine the mayonnaise, cream and curry powder in a bowl and season to taste with salt and pepper.

Add the pasta shells, chicken, ham or salami, celery and green pepper and mix well. Serve on a bed of lettuce leaves.

*Note:*
Any leftover cooked poultry can be used instead of chicken.

## GOOSE WITH HORSERADISH SAUCE

Serves 6

*1 kg cooked goose, cut into bite-sized pieces*
*200 ml bought horseradish sauce*
*1 small firm cabbage*
*125 ml vinaigrette dressing*

Place the goose in a bowl and stir in the horseradish sauce. Cover the bowl and set it aside in a cool place for 1 hour.

Meanwhile, remove the leaves from the cabbage and cut out the coarse ribs. Shred the cabbage very finely and bring it to the boil in a large saucepan of boiling salted water.

Cook for about 10 minutes, then drain through a colander and refresh the cabbage under cold running water. Drain well. When the cabbage is quite dry, mix it with the vinaigrette dressing.

Pile the goose in the centre of a serving platter and spoon the cabbage around the edges.

Serve with buttered rye bread.

*Note:*
Leftover cold turkey or duck may be used instead of goose.

## PINEAPPLE AND CHICKEN SALAD

Serves 4

*250 g cooked chicken, cut into strips*
*3 cherry tomatoes, quartered*
*¼ English cucumber, cut into thin strips*
*salt*
*100 g button mushrooms, sliced*
*fresh lemon juice*
*2 slices pineapple, chopped*
*butter lettuce leaves*

**Curry dressing**
*100 ml mayonnaise*
*30 ml fresh lemon juice*
*15 ml tomato purée*
*5 ml paprika*
*2 ml curry powder*
*2 ml ground ginger*
*salt and freshly ground black pepper*

Place the chicken in a bowl. Add the tomatoes and cucumber and sprinkle them lightly with salt. Sprinkle the mushrooms with lemon juice and add to the bowl. Add the pineapple chunks and toss the salad lightly.

Line a large salad bowl, or 4 individual bowls, with butter lettuce leaves and spoon the salad onto them.

Combine all the dressing ingredients well in a screw-topped jar and spoon over the salad.

Serve with Melba toast.

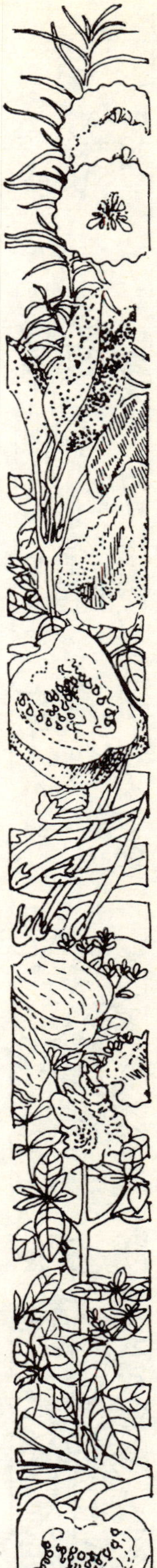

## MARINATED CHICKEN, BABY MARROW AND CAULIFLOWER SALAD

Serves 6

8 *baby marrows, trimmed and sliced*
½ *small cauliflower, broken into florets*
*250 ml coarsely chopped cooked chicken*
1 *medium onion, thinly sliced*
10 *button mushrooms*
*young spinach leaves*

**Marinade**

*250 ml olive oil*
*50 ml fresh lemon juice*
*50 ml fresh unsweetened orange juice*
1 *bay leaf, halved*
5 *black peppercorns*
*10 ml finely chopped fresh basil*
*10 ml finely chopped fresh oregano*
*salt*

**Garnish**

*bean sprouts or watercress sprigs*

Blanch the baby marrow slices and cauliflower florets in lightly salted water until barely tender, about 5 minutes.

Combine with the chicken, onion and mushrooms in a ceramic dish.

Combine the marinade ingredients and pour over the chicken mixture. Marinate in the refrigerator for at least 6 hours, stirring occasionally.

To serve, drain off the marinade and place the chicken and vegetables on young spinach leaves. Drizzle a little of the marinade over and garnish with bean sprouts or watercress sprigs.

## CHICKEN-STUFFED BAKED POTATOES

Serves 4

4 *large potatoes, scrubbed well and dried*

**Chicken filling**

15 ml *butter or margarine*
5 ml *salt*
2 ml *freshly ground black pepper*
1 ml *grated nutmeg*
50 ml *thick cream*
125 ml *very finely chopped cooked chicken*

**Garnish**

*dill sprigs*

Prick the potatoes lightly with a fork and bake them on the centre shelf of the oven at 190 °C for 1½ hours.

Remove the potatoes from the oven. Cut a cross in the top of each and squeeze gently to open.

Scoop out most of the flesh, leaving a 1 cm thick shell, and place flesh in a bowl.

Add the butter or margarine and mash well. Add the salt, pepper and nutmeg, mixing well.

Stir in the cream and beat until well blended, then stir in the chicken.

Stuff each potato shell with filling and place in a baking dish. Bake at 190 °C for 10 minutes, or until filling is lightly browned.

Garnish with dill.

Serve immediately, with salad.

*Note:*

It will cut the cooking time considerably if the potatoes are first cooked in a microwave oven. Prick them with a skewer and microwave on HIGH for 6-8 minutes, then proceed with the recipe.

*Right: Chicken and apple salad (p. 44)*

# TACOS WITH CHICKEN AND CHEESE FILLING

Serves 6

**Tacos**

300 g *cake flour*
100 g *fine mealie meal*
60 g *coarse yellow mealie meal*
10 ml *baking powder*
5 ml *salt*
100 g *margarine*
250 ml *water*

**Filling**

375 ml *shredded cooked chicken*
*one small head of lettuce, shredded*
*onion rings*
1 *large ripe tomato, skinned and chopped*
250 ml *grated Cheddar cheese*

First make the tacos. Sift the dry ingredients, cut in the margarine and rub until the mixture resembles coarse crumbs.

Add the water and mix to a soft dough. Knead until smooth and elastic on a lightly floured surface, about 8 minutes.

Shape the dough into 12 balls, cover and leave for 15 minutes.

Roll each ball out into a 15 cm diameter circle on a lightly floured surface.

Heat an ungreased heavy-based frying pan or griddle and cook tacos on each side for about 5 minutes, or until lightly browned. Allow to cool slightly; tacos should still be pliable.

Fold over to a U-shape and allow to dry.

Fill tacos with shredded chicken, lettuce, onion rings and tomato and sprinkle grated cheese on top.

Serve as a snack or light meal.

*Notes:*

Any other leftover cooked poultry may be used instead of chicken.

Taco shells may be bought ready-made at some supermarkets.

*Left: Tacos with chicken and cheese filling*

# TURKEY-STUFFED PITA BREADS

Serves 4

**Pita breads**

*180-250 ml warm water*
*15 ml dried yeast*
*15 ml sugar*
*625 ml bread flour*
*5 ml salt*
*5 ml sunflower oil*

**Turkey and avocado filling**

*250 ml chopped cooked turkey*
*1 large ripe avocado, peeled, stoned and chopped*
*10 ml fresh lemon juice*
*15 ml mayonnaise*
*1 large tomato, finely chopped*
*4 spring onions, finely chopped*
*2 stalks celery, chopped*
*75 ml grated Cheddar cheese*
*salt and freshly ground black pepper*

**Garnish**

*shredded lettuce*
*bean sprouts*

First make the pita breads. Combine 60 ml warm water with yeast and sugar and stir to dissolve.

Leave in a warm place for 5 minutes or until bubbly.

Sift flour and salt into a mixing bowl, add the oil and then the yeast mixture and beat until well mixed.

Add enough water to make dough form a ball.

Turn out onto a lightly floured board and knead until smooth and elastic, 5-8 minutes.

Allow to rise in a warm place, covered, until doubled in bulk, 45 minutes to 1 hour.

Punch down, divide into 12 pieces and roll each into a ball.

Roll each ball out very thinly on a lightly floured surface to a circle 15 cm in diameter. Place on lightly greased baking sheets and leave in a warm place, covered, for 15 minutes.

Bake at 200 °C for 7-8 minutes, or until puffed up and golden. Remove to a wire rack to cool.

To make the filling, combine all the ingredients and season to taste with salt and pepper.

To serve, cut 4 pita breads in half and carefully split each half open to form a pocket. Add the filling and top with shredded lettuce and bean sprouts.

*Note:*
The leftover pita breads may be frozen. To serve, place frozen breads in a hot oven (200 °C) for a few minutes to crisp.

## RÖSTI WITH CHICKEN AND CHEESE TOPPING

Serves 4

**Rösti**
*750 g potatoes*
*45 ml sunflower oil*
*1 medium onion, grated*
*2 ml salt*
*2 ml freshly ground black pepper*

**Chicken and cheese topping**
*250 ml chopped leftover cooked chicken*
*125 ml grated Cheddar cheese*
*paprika*

Boil the potatoes in their jackets in lightly salted water for 10 minutes. Allow them to cool, then peel and grate them coarsely.

Heat the oil in a frying pan and sauté the onion until transparent, about 5 minutes.

Add the potatoes, salt and pepper and spread evenly to form a thick cake (rösti).

Fry over low heat for about 10 minutes, lifting the rösti from time to time to check that it does not become too brown underneath.

Lift the rösti out of the pan onto a plate.

Add a little more oil to the pan, if necessary, and return the rösti to it, cooked side uppermost.

Fry over low heat for a further 10 minutes, checking occasionally to make sure that the rösti does not get too brown underneath.

Top the rösti first with chicken and then with cheese. Sprinkle with paprika and place under a heated grill until the cheese bubbles.

Serve immediately, sliced, with a green salad.

## CHICKEN MAYONNAISE SNACKWICH

Serves 4

8 *slices bread*
*butter*
250 g *cooked chicken, coarsely chopped*
125 ml *peas*
60 ml *mayonnaise*
15 ml *chopped fresh mixed herbs*
*salt and freshly ground black pepper*

Butter the bread on both sides. Combine the chicken with the remaining ingredients and place ¼ of the mixture on each of 4 slices of bread.

Cover with the remaining slices of bread and place 2 sandwiches in the snackwich. Toast the sandwiches and remove them from the snackwich.

Repeat for the other 2 sandwiches and serve hot.

## SAVOURY TOASTS

*These savoury toasts are ideal as snacks or a light meal, and are a great way of using up leftovers*

Serves 4

4 *slices white or whole-wheat toast*
*topping of your choice**

## TURKEY AND BEANS TOPPING

250 ml *finely chopped leftover cooked turkey*
125 ml *baked beans in tomato sauce*
8 *thin slices mozzarella or mature Cheddar cheese*
*paprika*

Layer the turkey and beans on the slices of toast and cover with the cheese.

Sprinkle lightly with paprika and place under the grill until heated through and the cheese is bubbling.

Serve immediately.

## SAUTÉED CHICKEN, ONION AND MUSHROOM TOPPING

| | |
|---|---|
| *1* | *medium onion, diced* |
| *6* | *large mushrooms, chopped* |
| *15 ml* | *sunflower oil* |
| *250 ml* | *minced or very finely chopped cooked chicken* |
| | *salt and freshly ground black pepper* |
| | *butter or margarine* |

Sauté the onion and mushrooms in heated oil until softened, 5-7 minutes.

Add the chicken and cook until heated through, about 5 minutes, stirring occasionally.

Season to taste with salt and pepper and spoon onto toast.

Dot with butter or margarine and place under the grill until lightly browned.

Serve immediately.

## CHICKEN RAREBIT

| | |
|---|---|
| *30 ml* | *butter or margarine* |
| *30 ml* | *cake flour* |
| *250 ml* | *milk* |
| *250 ml* | *grated Cheddar cheese* |
| *250 ml* | *finely chopped cooked chicken* |
| *15 ml* | *Worcester sauce* |
| *5 ml* | *prepared French mustard* |
| | *salt and freshly ground black pepper to taste* |

Melt the butter or margarine in a saucepan over low heat.

Add the flour and cook, stirring constantly, for 2-3 minutes.

Gradually add the milk, stirring constantly until the sauce is smooth and thick.

Add most of the cheese, the chicken, Worcester sauce, mustard, salt and pepper to taste and simmer, stirring, until the cheese has melted.

Spread on the slices of toast and place under the grill until the cheese is bubbling.

Serve immediately.

## IDEAS FOR SANDWICH FILLINGS AND TOPPINGS

*Savoury butters for open sandwiches*
Cream 125 g softened butter with:
10 ml mustard powder and 5 ml prepared mustard.

10 ml finely chopped onion and 10 ml finely chopped green pepper.

10 ml finely chopped skinned and seeded tomato and 2 ml dried basil.

1 clove garlic, crushed, and 10 ml minced fresh parsley.

10 ml minced onion and 2 ml crumbled dried rosemary.

*Toppings for open sandwiches*
Slices of roast chicken topped with a tomato slice or green pepper ring and a slice of stuffed olive in the centre. Use tomato or green pepper butter.

Chicken liver pâté topped with a thin slice of English cucumber and a slice of radish.

Finely chopped cooked turkey combined with sour cream and very finely chopped onion. Top with a thin slice of tomato. Use garlic butter.

Sautéed chopped mushrooms, onion and shredded cooked chicken or duck topped with an onion ring. Use green pepper or onion and rosemary butter.

Thinly sliced roast duck topped with a little natural yoghurt and a green pepper ring. Use mustard butter.

*Sandwhich fillings*
Combine diced roast chicken, 5 ml finely chopped green pepper and enough mayonnaise to bind.

Place a lettuce leaf on the bread, top with 2-3 slices salami and thinly sliced cooked chicken, sliced tomato and an onion ring.

Chop cooked turkey or duck finely and combine with a little sour cream, finely chopped onion and mushrooms.

Slice cooked chicken and top with grated onion and a dab of fruit chutney.

Place sliced cooked turkey or chicken on a bed of lettuce and top with mashed blue cheese mixed with a little mayonnaise.

Slice cooked chicken or duck and top with sliced tomato, sliced mushrooms, a dab of mayonnaise or ricotta cheese and crumbled crisp bacon.

Mince cooked chicken or turkey and mix with ground almonds, natural yoghurt and chopped fresh parsley or watercress leaves.

## CHICKEN CANNELLONI

*A tasty version of the classic Italian dish – the chicken goes particularly well with the ricotta cheese and spinach*

Serves 4

*8 cannelloni tubes or lasagne sheets*

**Chicken and ricotta stuffing**
*250 ml finely chopped cooked chicken*
*250 g ricotta or chunky cottage cheese*
*250 g chopped cooked spinach, well drained*
*6 slices salami, chopped*
*30 ml grated Parmesan cheese*
*salt and freshly ground white pepper*

**Parmesan sauce**
*30 ml butter*
*30 g cake flour*
*300 ml milk*
*30 ml grated Parmesan cheese*
*salt and freshly ground white pepper*

**Garnish**
*30 ml grated Parmesan cheese*

Cook the cannelloni tubes or lasagne sheets in boiling salted water until just tender, about 15 minutes. Drain well, rinse and dry.

Combine all the stuffing ingredients well and season to taste with salt and pepper.

Stuff the cannelloni tubes, or place stuffing on the lasagne sheets and roll up, then place, seam down, in a greased oven-proof dish.

To make the sauce, melt the butter in a small saucepan and stir in first the flour and then the milk. Bring to the boil, stirring, and cook for 2-3 minutes, or until thickened. Stir all the time. Add the Parmesan cheese and season to taste with salt and pepper.

Pour the sauce over the cannelloni, sprinkle Parmesan cheese over and bake at 190 °C for 25-30 minutes or until cooked and lightly browned on top.

Serve hot with a mixed salad.

*Notes:*

Pancakes may be used if cannelloni tubes or lasagne sheets are unavailable.

Any other leftover cooked poultry may be used instead of chicken.

# BRIOCHES STUFFED WITH CHICKEN LIVERS AND EGG

Serves 6

**Brioches**

*625 ml cake flour*
*5 ml dried yeast*
*20 ml castor sugar*
*65 ml warm milk*
*pinch salt*
*125 ml butter*
*3 eggs, lightly beaten*
*1 egg yolk*
*15 ml milk*

**Chicken liver and egg stuffing**

*30 ml butter*
*100 g chicken livers*
*50 g cream cheese or ricotta cheese*
*2 ml salt*
*50 ml cream*
*3 eggs*
*10 ml chopped chives*

First make the brioches. Place 125 ml flour in a warm bowl. Dissolve the yeast with 5 ml of the sugar in the warm milk. Strain the yeast mixture into the flour and mix together. Place, covered with a damp cloth, in a warm place for 10-15 minutes or until frothy.

Place the remaining 500 ml flour and the salt in a mixing bowl. Rub in the butter and add the remaining sugar, the yeast mixture and the lightly beaten eggs. Mix well to form a soft batter.

Cover and leave in a warm place until doubled in bulk, about 1 hour. Punch down and knead lightly in the bowl. If the dough is too soft to handle, refrigerate it for a while.

Place rounds of dough in greased muffin tins. Make a funnel in the top of each roll and dampen it. Place a knob of dough in the funnel. Set aside and allow to rise until doubled in bulk, about 15 minutes.

Beat the egg yolk with the milk and use to brush the brioches.

Bake at 210 °C for 10-15 minutes, or until golden.

Cut the tops from the brioches and scoop out a little of the bread to leave room for the stuffing.

To make the stuffing, melt 15 ml of the butter in a frying pan and sauté the chicken livers in it until just cooked – about 8 minutes. Chop the livers coarsely and remove them from the pan.

Melt the cream cheese or ricotta cheese and the remaining butter in the pan. Beat the salt, cream and eggs together lightly and pour into the melted cheese and butter.

Cook, stirring, until just beginning to set, then add the livers and sprinkle with chives.

Fill the brioches and serve immediately with a green salad.

*Note:*
Freeze the leftover brioches. To serve, place them in a cooking bag and heat at 180 °C for 15-20 minutes.

## CHICKEN AND MUSHROOM QUICHE

Serves 6

*250 g short crust pastry**

**Filling**

*6 large mushrooms, thinly sliced*
*2 eggs*
*1 egg white*
*100 ml milk*
*60 ml cream*
*2 ml salt*
*2 ml freshly ground black pepper*
*250 ml minced cooked chicken*
*15 ml chopped fresh parsley*

Press the pastry into a well-greased 20 cm diameter quiche pan.

Layer the mushrooms in the pastry shell.

Beat the eggs, egg white, milk and cream, then stir in the salt, pepper, chicken and parsley. Pour into the pastry shell and bake at 220 °C for 10 minutes.

Lower the temperature to 180 °C and bake for a further 10 minutes. Serve hot, with a mixed salad.

*Note:*
Any leftover cooked poultry may be used instead of chicken.

## CHICKEN WITH ASPIC

*The classic chaudfroid of French cooking brought right up to date – less fuss but all the flavour of the original*

Serves 6

1 ℓ *water*
2 *chicken stock cubes, crumbled*
1 *stalk celery, chopped*
5 ml *peppercorns*
1 *bay leaf*
6 *chicken breasts*

**Sauce**

10 ml *gelatine powder*
250 ml *mayonnaise*
15 ml *fresh lemon juice*
250 ml *thick cream*

**Aspic jelly**

250 ml *water*
60 ml *dry sherry*
15 ml *gelatine powder*

**Garnish**

*watercress sprigs or fennel tips*

Combine the water, stock cubes, celery and spices in a saucepan. Bring to the boil while stirring, then reduce the heat and poach the chicken, covered, for 10 minutes or until just done. Remove the chicken from the saucepan and drain well. Leave to cool, then refrigerate for 30 minutes.

Strain the stock through a fine-mesh sieve and reserve 250 ml.

To make the sauce, sprinkle the gelatine over the reserved stock and dissolve over hot water. Allow to cool slightly, then add the remaining ingredients.

Coat the chicken with some of the sauce and refrigerate for 5 minutes. Repeat until all the sauce has been used.

To make the aspic, combine the water, sherry and gelatine in a bowl, then stir until dissolved over hot water. Pour into a flat dish and refrigerate until set.

Cut the aspic into chunks and spread out on a serving platter. Place the chicken on top and garnish with watercress or fennel.

Serve with spinach and walnut salad.

## COLD CHICKEN WITH CURRY MAYONNAISE

Serves 4-6

| | |
|---|---|
| *1 x 1,5 kg* | *chicken, cut into portions* |
| | *salt and freshly ground white pepper* |
| | *sunflower oil* |

**Curry mayonnaise**

| | |
|---|---|
| *15 ml* | *olive oil* |
| *1* | *medium onion, finely chopped* |
| *10 ml* | *curry powder* |
| *5 ml* | *tomato purée* |
| *125 ml* | *dry red wine* |
| *100 ml* | *water* |
| *1* | *thin slice lemon rind* |
| *1* | *bay leaf* |
| *15-30 ml* | *apricot jam* |
| *1 ml* | *Tabasco sauce* |
| | *salt and freshly ground white pepper* |
| | *sugar* |
| *125 ml* | *mayonnaise* |

**Garnish**

*lettuce leaves*
*sliced ripe tomatoes*
*radishes*

Season the chicken portions with salt and pepper and brush lightly with sunflower oil. Grill, turning often, until the chicken is cooked and lightly golden – 20-30 minutes. Drain and allow to cool.

To make the curry mayonnaise, heat the oil in a saucepan and sauté the onion until softened, about 5 minutes. Add the curry powder and cook for 1 minute. Add the tomato purée, wine, water, lemon rind and bay leaf.

Bring to the boil and cook for about 10 minutes, stirring occasionally. Stir in the apricot jam, season with Tabasco sauce, salt and pepper and sugar to taste.

Strain the sauce and adjust the seasoning if necessary. Stir 60 ml of the sauce into the mayonnaise.

Serve the chicken on a bed of lettuce leaves, garnished with sliced tomatoes and whole trimmed radishes. Serve the curry mayonnaise separately.

*Note:*

As the curry sauce keeps well in the refrigerator, it is worth making more than is immediately needed and keeping it for adding to mayonnaise, cream or yoghurt when required.

## CHICKEN AND NUT ROULADE

Serves 6-8

**Roulade**

*60 ml butter or margarine*
*125 ml cake flour*
*500 ml milk*
*5 eggs, separated*
*salt and freshly ground white pepper*
*pinch sugar*

**Chicken and nut filling**

*200 g cream cheese or smooth cottage cheese*
*2 stalks celery, very finely chopped*
*250 g cooked chicken, very finely chopped*
*50 g walnuts, very finely chopped*
*mayonnaise*
*salt and freshly ground white pepper*

**Garnish**

*watercress sprigs*

First make the roulade. Melt the butter or margarine, add the flour and cook, while stirring, for 1 minute.

Blend in the milk and bring to the boil, stirring constantly. Remove from the heat and beat in the egg yolks with seasoning and sugar to taste.

Whisk the egg whites until stiff and fold into the flour mixture.

Spread the mixture evenly into a greased and lined 23 x 33 cm swiss roll tin and bake for 30-35 minutes at 180 °C, or until golden brown.

Turn out onto a damp cloth, remove the lining paper and roll up the roulade, using the cloth as a guide.

To make the filling, combine the cheese, celery, chicken and walnuts with mayonnaise and seasoning to taste.

Unroll the roulade and spread the filling mixture over it. Roll it up again and place on a serving platter.

Garnish with watercress and serve hot or cold, sliced, with a green salad.

# CHICKEN PIZZA

Serves 4

**Dough**

*500 ml bread flour*
*1 ml salt*
*15 ml instant yeast*
*2 ml sugar*
*200 ml lukewarm milk or water*
*30 ml sunflower oil*

**Topping**

*15 ml olive oil*
*4 tomatoes, skinned and sliced*
*2 ml freshly ground black pepper*
*5 ml fresh oregano*
*100 g leftover roast chicken, finely chopped*
*125 g salami, thinly sliced*
*150 ml grated Cheddar cheese*

**Garnish**

*capers or black olives*

First make the dough. Sift the flour and salt into a bowl, add the yeast and sugar.

Add the milk or water and mix well. Knead the oil into the dough and knead for 5 minutes.

Cover and leave to rise in a warm place for 30 minutes, or until doubled in bulk.

Knock back the dough and then roll it out thinly on a lightly floured surface. (For smaller pizzas, divide into two before rolling out.)

Place on a greased baking sheet and turn up the edges slightly to form a rim.

Brush the dough with a little olive oil.

Arrange sliced tomatoes on top and sprinkle with pepper and oregano. Top with the chicken and salami, then the grated cheese, and garnish with capers or olives.

Leave for 10 minutes, then bake at 200 °C for 20 minutes.

Serve hot, accompanied by a large mixed salad.

*Notes:*

Pizzas may also be baked in a Weber braai, with the lid on.

For a light snack, arrange the topping on 4 slices of whole-wheat or white bread and brown under the grill until the cheese bubbles.

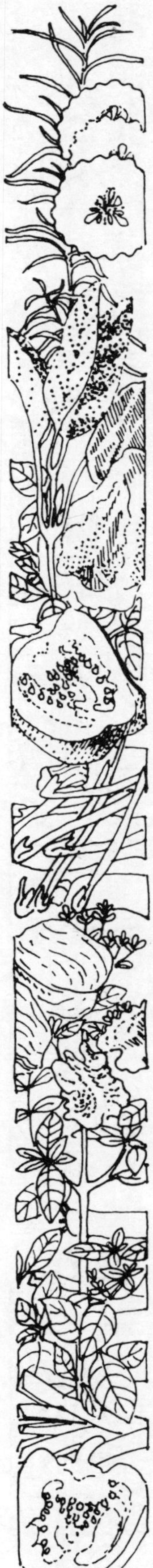

## CHINESE STIR-FRIED CHICKEN

*An ideal way to use up leftover chicken and vegetables*

Serves 6

*15 ml sunflower oil*
*6 chicken breast fillets, cubed*
*1 red sweet pepper, seeded and cut into strips*
*1 stalk celery, chopped*
*4 water chestnuts, chopped*
*100 g small button mushrooms*
*100 g mangetouts or young peas in their pods*

**Marinade**

*30 ml chicken stock**
*30 ml sunflower oil*
*15 ml soy sauce*
*5 ml ginger syrup*

**Garnish**

*50 ml chopped cashew nuts*

First make the marinade. Combine all the ingredients in a large bowl and marinate the chicken cubes for 1-2 hours. Drain.

Heat a wok or heavy-based frying pan to very warm, add 10 ml oil and stir-fry the chicken for 4-5 minutes. Remove the chicken from the frying pan.

Heat a little more oil in the frying pan and stir-fry all the vegetables for 2-3 minutes.

Add the chicken and stir-fry for a further minute. Sprinkle the cashew nuts over.

Serve immediately with rice and a little extra soy sauce.

*Note:*
The vegetables must still be crisp.

## CHICKEN SOUFFLÉ

Serves 4

*30 ml butter or margarine*
*30 ml cake flour*
*300 ml warm milk*
*250 ml grated Cheddar cheese*
*250 ml minced cooked chicken*
*2 ml salt*
*1 ml freshly ground white pepper*
*5 eggs, separated*

Melt the butter or margarine in a saucepan and stir in the flour until smooth. Add the milk and bring to the boil. Cook, stirring constantly, until thickened – about 5 minutes.

Add the cheese, chicken, salt and pepper and remove from the stove.

Beat the egg yolks well and add to the mixture.

Whisk the egg whites until stiff but not dry and fold into the mixture.

Pour into a 1,5 ℓ soufflé dish and bake at 180 °C for 30-40 minutes, or until well risen and lightly golden.

Serve immediately with a green salad and Melba toast.

*Note:*
Any leftover cooked poultry may be used instead of chicken.

## FARMER'S OMELETTE

Serves 4

*30 ml butter or margarine*
*1 small onion, finely chopped*
*½ small green pepper, seeded and finely chopped*
*60 g mushrooms, thinly sliced*
*60 g rindless streaky bacon, chopped*
*3 eggs*
*75 ml milk*
*5 ml dried mixed herbs*
*5 ml salt*
*2 ml freshly ground white pepper*
*100 g leftover cooked chicken, diced*
*2 ripe tomatoes, skinned and coarsely chopped*
*100 g mature Cheddar cheese, grated*

Melt the butter or margarine in a large frying pan and sauté the onion, green pepper, mushrooms and bacon in it until soft – about 5 minutes.

Meanwhile, beat the eggs, milk, herbs, salt and pepper thoroughly in a bowl.

Add the egg mixture to the pan and sprinkle with the chicken, tomatoes and cheese.

Fry over medium heat until the base is firm and the eggs lightly set.

Serve immediately with crusty French bread and a salad.

## CHICKEN CROQUETTES

Serves 4

*250 ml Béchamel* or white sauce**
*30 ml butter or margarine*
*1 large onion, finely chopped*
*4 rashers rindless bacon, finely chopped*
*250 ml leftover cooked chicken*
*15 ml finely chopped fresh parsley*
*1 egg, lightly beaten*
*fine dry bread crumbs*
*sunflower oil*
*fresh lemon juice*

Make a thick Béchamel or white sauce and let it cool completely.

Heat butter or margarine in a frying pan and sauté the onion and bacon until the onion is golden brown, about 5 minutes.

Combine the onion mixture, chicken, parsley and cold sauce and form into small croquettes.

Dip the croquettes first in beaten egg and then in the bread crumbs.

Deep-fry the croquettes in hot oil until golden brown, 5-7 minutes. Drain on absorbent paper and sprinkle with lemon juice.

Serve at once with minted new potatoes and sliced ripe tomatoes.

## CRUNCHY ASPARAGUS CHICKEN

Serves 4

*250 g leftover cooked chicken, chopped*
*1 x 410 g can asparagus cuts, drained*
*50 g pecan nuts, coarsely chopped*
*250 g cheese sauce**
*250 ml grated Cheddar cheese*
*1 x 50 g packet plain potato crisps, crumbled*

Layer the chicken in a greased ovenproof dish and sprinkle the asparagus and nuts over.

Pour the cheese sauce over and top with grated cheese.

Sprinkle the crisps on top, covering the whole surface, and bake at 180 °C until heated through and the cheese is bubbling – about 20 minutes.

Serve hot, with a crisp mixed salad and whole-wheat toast.

*Right: Rösti with chicken and cheese topping (p. 51)*

## CHICKEN WITH TUNA

*The Italians usually make this dish with veal and tuna, but the chicken makes a tasty change*

Serves 6-8

*8 chicken breast fillets*
*1 carrot, finely chopped*
*1 onion, chopped*
*1 bay leaf*
*small bunch parsley*
*pinch dried thyme*
*celery salt*
*freshly ground black pepper*
*cold water*

**Tuna sauce**

*1 x 200 g can tuna, drained*
*300 ml mayonnaise*
*4 anchovy fillets*
*45 ml fresh lemon juice*
*30 ml capers*

**Garnish**

*freshly chopped parsley*
*lemon slices*

Arrange the chicken in a shallow saucepan. Add the carrot, onion, bay leaf, parsley, thyme, celery salt and pepper to taste.

Just cover with water and bring slowly to the boil. Simmer very gently until chicken is just tender, about 20 minutes. Cool in the stock, then remove the chicken. Slice the chicken thinly.

To make the sauce, mash the tuna and blend it with the mayonnaise, anchovies, lemon juice and capers in a blender or food processor until smooth.

If the sauce seems too thick, thin it down with a little of the chicken stock and adjust the seasoning, if necessary.

Spread some of the sauce over a serving platter and top with some of the chicken slices.

Continue layering in this manner until all the ingredients have been used, ending with a layer of sauce. Cover and refrigerate overnight.

Garnish with parsley and lemon slices.

Serve with chilled cooked julienne vegetables.

*Left: Chicken with aspic (p. 58)*

## CRÊPES WITH CHICKEN AND ARTICHOKE FILLING

Serves 6-8

**Crêpes**

*200 ml cake flour*
*3 eggs, lightly beaten*
*15 ml sunflower oil*
*250 ml milk*
*butter*

**Chicken and artichoke filling**

*30 ml butter*
*1 onion, thinly sliced*
*4 rashers rindless bacon, chopped*
*150 ml sour cream*
*30 ml brandy*
*½ chicken stock cube*
*10 ml cornflour*
*125 ml water*
*500 ml chopped cooked chicken*
*1 x 410 g can artichoke hearts, drained and chopped*

**Topping**

*30 ml butter*
*250 ml fresh bread crumbs*
*250 ml grated Cheddar cheese*
*30 ml chopped fresh parsley*

First make the crêpes. Sift the flour into a bowl and make a well in the centre. Combine the eggs, oil and milk and pour into the well in the flour. Gradually whisk the liquid into the flour to make a smooth batter. Strain the batter through a sieve into a jug to make pouring easier. Set aside for about 1 hour.

Heat a little butter in a frying pan over medium heat and swirl it all over the pan. Pour 30-45 ml batter into the pan and swirl it around to cover the base completely. Brown the crêpe lightly underneath, about 2 minutes (lift it with a spatula to check), and turn it over to brown the other side. Slide the crêpe onto a plate and keep it warm while making the other crêpes.

To make the filling, heat the butter in a frying pan and stir the onion and bacon over medium heat until the onion is transparent – about 5 minutes. Stir in the sour cream, brandy, crumbled stock cube and cornflour blended to a paste with water.

Stir until the mixture boils and thickens, then reduce the temperature, stir in the chicken and artichokes and stir for a further minute.

Place 30 ml filling on each crêpe and roll it up. Arrange the crêpes in a single layer in a greased ovenproof dish.

To make the topping, melt the butter in a frying pan and stir in the bread crumbs. Continue stirring until the bread crumbs are golden, then stir in the cheese and parsley and spoon the mixture over the crêpes.

Bake, uncovered, at 180 °C for 15 minutes.

Serve with buttered baby vegetables or a mixed salad.

*Note:*
Mushrooms may be used instead of the artichokes. Sauté 250 g chopped mushrooms with the onion and bacon.

## GALANTINE OF DUCK

Serves 6

*1 x 2 kg duck, boned*
*sunflower oil*

**Chicken and ham stuffing**
*750 g minced chicken*
*2 egg yolks, lightly beaten*
*30 ml chopped fresh parsley*
*30 ml tomato paste*
*250 ml day-old bread crumbs*
*15 ml currants*
*500 g ham, cut into slivers*

Turn the flesh of the legs and wings to the inside of the boned duck and lay the duck down flat, skin underneath.

To make the stuffing, combine the chicken mince, egg yolks, parsley, tomato paste, bread crumbs and currants in a large bowl and mix well. Spoon half the mixture down the centre of the duck and arrange half the ham slivers on top. Cover with the remaining stuffing and then the remaining ham slivers.

Fold first one and then the other side of the duck over the breadcrumb mixture and sew the flesh together with a needle and dark thread. Tie with string at 5 cm intervals to keep the galantine in shape during cooking.

Place the galantine in a roasting pan and rub it with oil. Bake at 180 °C for 1½ hours, or until the duck is tender. Allow to stand for 15 minutes before removing the string and thread.

Serve hot with buttered new potatoes, peas and carrots or cold, with a salad of oak lettuce, olives and kumquats.

## GREEN PEPPERS STUFFED WITH CHICKEN AND RICE

Serves 6

6 *medium green peppers, tops cut off and seeded*

**Chicken and rice filling**

1 *small onion, very finely chopped*
50 ml *olive oil*
150 ml *cooked brown rice*
150 ml *finely chopped cooked chicken*
30 ml *finely chopped parsley*
1 ml *dried thyme*
1 ml *dried sage*
500 ml *tomato purée*

First make the filling. Sauté the onion in a little of the olive oil until transparent, about 5 minutes.

Combine with the other stuffing ingredients, except the remaining olive oil and the tomato purée, and fill the peppers with the stuffing.

Place the peppers in a greased casserole dish.

Combine the remaining olive oil and the tomato purée and pour over the green peppers.

Bake at 180 °C for 45 minutes.

Serve hot, with sautéed potatoes and steamed peas.

## GUINEA FOWL RISOTTO

Serves 4-6

5 ml *sunflower oil*
1 *large onion, finely chopped*
2 *cloves garlic, crushed*
200 g *brown mushrooms, sliced*
1 *green pepper, seeded and thinly sliced*
250 g *uncooked brown rice*
150 ml *dry red wine or dry sherry*
100 g *sultanas*
100 g *plump pitted prunes, coarsely chopped*
750 ml *chicken stock**
*salt and freshly ground black pepper*
250 g *leftover roast guinea fowl, cut into strips*
15 ml *chopped fresh parsley*

Heat the oil in a large frying pan and sauté the onion, garlic and mushrooms over low heat for 10 minutes, stirring occasionally. Do not allow to brown.

Add the green pepper and rice and stir for 1 minute, then add the wine, sultanas, prunes and stock, season with salt and pepper and stir well. Simmer, covered, over medium heat for about 40 minutes, or until all the liquid has been absorbed and the rice is tender.

Add the guinea fowl and cook for 5-10 minutes, or until heated through, then stir in the parsley and transfer to a heated serving dish.

Serve with a green salad.

*Note:*
This is an ideal way to use up any leftover poultry. If using turkey, substitute dried apricots for prunes.

## SPAGHETTI WITH GREEN CHICKEN SAUCE

Serves 4-6

*400 g spaghetti*

**Green chicken sauce**
*500 g broccoli, lightly steamed and puréed*
*175 g cooked chicken, minced*
*150 ml natural yoghurt*
*salt and freshly ground black pepper*
*1 clove garlic, crushed*
*15 ml chopped stoned dates*

**Garnish**
*15 ml chopped walnuts*

Boil the spaghetti in boiling salted water until just tender, 12-13 minutes. Drain and keep hot.

Combine the broccoli, chicken, yoghurt, salt and pepper to taste and garlic in a saucepan and just heat through. Do not allow to boil. Stir in the dates.

Transfer the spaghetti to a heated deep serving dish and top with the sauce. Toss well and sprinkle with the nuts.

Serve with a salad of butter lettuce, cherry tomatoes and sliced English cucumber.

## SPINACH AND DUCK TERRINE

Serves 6-8

*1 bunch spinach, coarse stalks removed*

**Duck filling**

*15 ml butter*
*1 onion, finely chopped*
*2 rashers rindless bacon, chopped*
*10 ml curry powder*
*4 cooked duck breast fillets, chopped*
*65 ml cream*
*2 eggs, lightly beaten*

**Spinach filling**

*1 bunch spinach, coarse stalks removed*
*15 ml butter*
*1 clove garlic, crushed*
*200 ml cooked rice*
*30 ml cream*
*1 egg, lightly beaten*
*1 ml grated nutmeg*

**Tomato sauce**

*3 ripe tomatoes, skinned and chopped*
*1 onion, chopped*
*15 ml red-wine vinegar*
*2 ml sugar*

First prepare the casing. Rinse the spinach and steam the leaves until tender. Drain well and use to line a greased 1,5 ℓ ovenproof dish, reserving a few for the top of the terrine.

To make the duck filling, melt the butter in a frying pan, add the onion, bacon and curry powder and cook, while stirring, until the onion is soft – about 5 minutes. Purée with the duck, cream and eggs until smooth.

To make the spinach filling, rinse the spinach and chop it finely. Melt the butter in a saucepan, add the garlic and spinach and stir over high heat until the spinach is wilted – about 5 minutes. Drain, pressing out as much moisture as possible, and combine with the rice, cream, egg and nutmeg in a bowl.

Spread half the duck filling in the spinach-lined dish and top with the spinach filling and then the remainder of the duck filling. Top with the reserved spinach leaves and cover the dish with foil.

Place the terrine in a baking dish with enough hot water to come halfway up the sides of the terrine. Bake at 180 °C for 1 hour, or until set.

Meanwhile, make the tomato sauce. Combine all the ingredients in a saucepan and bring to the boil. Reduce the temperature and simmer, covered, for 10 minutes, stirring occasionally. Purée until smooth, then strain through a fine-mesh sieve.

Remove the terrine from the water and allow to stand for 5 minutes. Remove the foil and invert the terrine onto a serving dish.

Reheat the tomato sauce and serve with the terrine.

*Notes:*

The terrine may be served either hot or cold, but does not reheat well.

Chicken or turkey may be used instead of duck.

## QUICK TURKEY TETRAZZINI

Serves 6

*350 g cooked turkey, cubed*
*5 ml finely chopped fresh mixed herbs*
*100 g cooked noodles*
*100 g sautéed mushrooms*
*100 ml dry bread crumbs*
*75 ml grated Parmesan cheese*

**Sauce**

*30 ml butter or margarine*
*45 ml cake flour*
*250 ml milk or thick cream*

To make the sauce: Melt the butter or margarine in a saucepan, stir in the flour and cook, stirring, for 1 minute to make a roux. Add the milk or cream and cook, stirring, over low heat until thickened.

Remove from the stove and add the turkey, herbs, noodles and mushrooms. Spoon the mixture into a casserole dish and sprinkle the bread crumbs and cheese over.

Bake at 190 °C for 10 minutes, or until the topping is golden.

Serve as a light supper with a green salad.

*Note:*

Any other leftover poultry may be used instead of turkey.

## CHICKEN FONDUE

Serves 4

*sunflower oil*
*600 g chicken, cubed*
*1 pineapple, peeled and cubed*
*1 French loaf, cubed*
*400 g small button mushrooms*
*sauce of your choice**

Heat the oil in a fondue pot and place over flame.

Spear chicken, pineapple, bread cubes and mushrooms with fondue forks and cook in the hot oil.

Dip into the sauce of your choice.

Serve with a green salad.

## SWEET-AND-SOUR SAUCE

Makes 400 ml

*30 ml vinegar*
*20 ml sugar*
*5 ml tomato purée*
*10 ml cornflour*
*7 ml soy sauce*
*300 ml pineapple canning liquid mixed with water*
*30 ml sunflower oil*
*85 g canned pineapple, finely chopped*
*60 g onions, finely chopped*

Combine the vinegar, sugar, tomato purée, cornflour and pineapple liquid in a saucepan and bring to the boil. Cook until thickened, about 10 minutes.

Add the oil and continue boiling for 1-2 minutes. Stir in the pineapple and onions, blending well. Serve hot.

*Note:*
Keep sauces hot on a hot tray or over a burner.

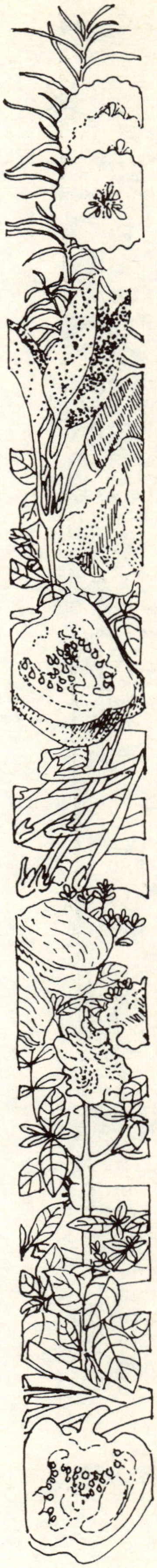

## FRUITY CURRY SAUCE

Makes 250 ml

*15 ml curry powder*
*15 ml cake flour*
*15 ml brown sugar*
*10 ml salt*
*2 ml mustard powder*
*2 bay leaves*
*4 peppercorns*
*1 clove garlic, crushed*
*15 ml vinegar*
*250 ml water*
*2 onions, thinly sliced*
*1 green pepper, seeded and thinly sliced*
*125 ml dried apricots*

Combine all the ingredients except the vinegar, water, onions, green pepper and apricots in a saucepan.

Add the vinegar and water and mix well. Stir in the onions, green pepper and apricots.

Simmer over medium heat, stirring, until the sauce thickens - about 5 minutes. Remove from the stove and cool.

## CHILLI SAUCE

Makes 400 ml

*30 ml butter*
*15 ml cornflour*
*1 medium onion, grated*
*1 small chilli, finely chopped*
*375 ml chicken stock**

Melt the butter in a saucepan, add the cornflour and stir until smooth.

Combine the onion, chilli and chicken stock in a small bowl and stir into the butter and cornflour mixture.

Simmer over low heat, stirring constantly, until thickened - about 5 minutes. Remove from the stove and serve either hot or cold.

# MAIN COURSE MEALS

This is where poultry really comes into its own. There are recipes for roasts, grills and braais; for fried dishes and braised ones; for stews, casseroles and baked meals . . . and each of them demonstrates anew how well poultry goes with a wide variety of vegetables, fruits, herbs and wines. You'll be able to choose recipes family and friends will love, such as Traditional chicken pie, as well as more elaborate ones for elegant entertaining, such as Chicken with sparkling wine or Port-marinated stuffed turkey. Some familiar dishes get a new look, for instance Chicken-stuffed cabbage rolls and Turkey fricadels with sour-cream sauce, while others will be new gastronomic experiences, such as Greek-style grilled chicken and Deep-fried chicken with walnuts. The choice is wide, and it's all yours.

## CHICKEN STUFFED WITH FIGS AND RICE

Serves 4

*1 x 1,5 kg chicken*
*olive oil*
*salt*
*cayenne pepper*

**Fig and rice stuffing**

*75 ml butter or margarine*
*2 medium onions, thinly sliced*
*chicken giblets*
*10 fresh figs, peeled and cut lengthwise into 8 slices*
*250 g uncooked rice*
*500 ml water*
*salt and freshly ground black pepper*

First make the stuffing. Heat 60 ml of the butter or margarine in a large deep frying pan over medium heat. Sauté the onions until transparent, about 5 minutes.

Chop the chicken giblets coarsely and add, with the figs, to the onions. Cook, stirring occasionally, until the giblets lose their pink colour.

Add the rice and stir until the grains are well coated with butter. Add water, salt to taste and a little pepper and bring to the boil. Reduce the heat and simmer gently for 25-30 minutes, or until all the liquid has been absorbed.

Remove the frying pan from the stove and add the remaining butter or margarine.

Dry the inside of the chicken well and stuff it with about ²/₃ of the fig mixture, reserving the rest. Sew up the chicken and place it, back uppermost, in a fairly deep roasting pan.

Smear the bird with oil, sprinkle with salt and cayenne pepper and place the pan on the middle shelf of the oven. Roast for 1½ hours at 180 °C. Remove the thread.

Reheat the reserved stuffing about 5 minutes before serving and serve with the carved chicken. Serve with a medley of young steamed vegetables.

## SMOKED CHICKEN

Serves 4

*1 x 1,5 kg chicken*
*5 ml salt*
*5 ml chicken stock powder*
*2 ml freshly ground black pepper*
*2 ml sugar*
*15 ml cake flour*

Trim the chicken, removing the wing tips and oil glands.

Combine the salt, stock powder, pepper, sugar and cake flour and season the chicken with the mixture, inside and out. Truss the chicken.

Sprinkle 30 ml untreated oak or other hardwood shavings in the base of a large deep heavy-based saucepan (preferably cast-iron) with a tight-fitting lid and stand a trivet in the saucepan.

Place the chicken in a metal container slightly smaller than the saucepan and stand it on the trivet. Close the lid tightly.

Turn the stove on high – if your stove has solid plates or a glass top, preheat it for 5 minutes – and smoke the chicken for 20 minutes.

Turn the temperature down to medium and continue smoking for 1 hour. Switch off the stove, leaving the saucepan on it for a further 15 minutes.

Serve either hot or cold.

*Note:*
To smoke a 2 kg turkey, salt it inside and out with 7 ml salt, then combine 7 ml salt, 10 ml chicken stock powder, 5 ml freshly ground black pepper, 5 ml sugar and 30 ml cake flour and season the turkey with the mixture inside and out. Sprinkle 30-45 ml untreated oak or other hardwood shavings into the saucepan and smoke the turkey as for the chicken.

## POT-ROASTED CHICKEN

Serves 4

*1 x 1,5 kg chicken*
*salt and freshly ground black pepper*
*several parsley sprigs or 1 clove garlic*
*100 g butter*
*300 ml chicken stock**
*10 ml cornflour*

Wipe the chicken with a damp cloth. Season the cavity with salt and pepper and place parsley sprigs or garlic and 30 ml butter inside.

Place the chicken in a flameproof casserole dish or a large heavy-based saucepan and brown it all over in 15 ml butter over medium heat - about 10 minutes. (Turn the chicken by using 2 spoons to prevent the skin splitting.)

Lower temperature, add the remaining butter and stock. Simmer, covered, until tender - about 1-1¼ hours - turning occasionally.

Transfer the chicken to a serving platter and keep it warm.

Thicken the pan juices with the cornflour mixed with a little water and stir over high heat to make a gravy.

Serve the chicken carved, with the gravy, potato croquettes and steamed young vegetables.

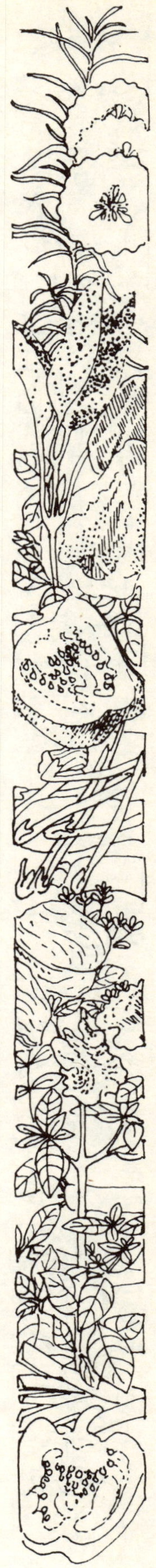

## CHICKEN WITH SPARKLING WINE

Serves 8

*2 x 1,5 kg chickens*
*salt and freshly ground black pepper*
*sunflower oil*
*150 ml dry sparkling wine*
*200 ml sour cream or natural yoghurt*
*chopped fresh tarragon*
*3 egg yolks*

**Garnish**

*chopped fresh tarragon or chervil*

Season the chickens inside and out with salt and pepper and rub the skin with oil. Roast at 200 °C for 1 hour, or until cooked and crisp, basting often with the pan juices. Quarter each chicken and keep warm and covered.

Skim off excess fat from the roasting pan and add the sparkling wine to the pan. Stir well and add the cream or yoghurt and chopped tarragon to taste.

Add a little of the hot sauce to the egg yolks and blend thoroughly. Return to the sauce and stir continuously over very low heat (or over a saucepan of boiling water) until thickened, about 5 minutes.

Pour the sauce over the chickens, garnish with tarragon or chervil and serve with steamed new potatoes and baby marrows.

## CORNISH HEN WITH LEMON AND HERBS

Serves 4

| | |
|---:|---|
| *1 x 2 kg* | *Cornish hen* |
| *1* | *lemon, quartered* |
| *15 ml* | *chopped fresh sage* |
| *15 ml* | *chopped fresh rosemary* |
| *125 ml* | *melted butter* |
| | *salt and freshly ground black pepper* |
| *2* | *cloves garlic, crushed* |

**Garnish**

*fresh rosemary sprigs*
*lemon slices*

Arrange the Cornish hen on the rack of a roasting pan and place the lemon quarters inside the cavity.

Combine the sage and rosemary and use half to stuff the cavity.

Pour the butter over the Cornish hen and sprinkle the remaining sage and rosemary mixture over it. Season with salt and pepper.

Roast, basting often with the pan juices, at 200 °C for about 1 hour.

Sprinkle the garlic over the hen and roast it for a further 5-10 minutes, or until the juices run clear when the fleshy part of a thigh is pricked with a fork.

Transfer the hen to a heated platter and garnish with rosemary sprigs and lemon slices.

Serve with rösti*, buttered petits pois and baby carrots.

## LEMON CHICKEN IN A CLAY POT

Serves 4

| | |
|---:|---|
| *1 x 1,5 kg* | *chicken* |
| *15 ml* | *olive oil* |
| *10 ml* | *salt* |
| *2 ml* | *freshly ground white pepper* |
| *1* | *bouquet garni* |
| *2* | *cloves garlic, crushed* |
| *1* | *small onion, chopped* |
| *2* | *ripe lemons, sliced* |

Submerge the clay pot in water for 30 minutes.

Rub the chicken with olive oil, salt and pepper and place the bouquet garni, garlic and onion inside the cavity. Truss the chicken.

Drain the clay pot and layer the lemon slices in the bottom. Place the chicken on top.

Cover with the lid and place in a cold oven. Set the temperature to 200 °C and bake for 1½ hours.

Serve with brown rice, steamed baby marrows and onions.

## DUCK À L'ORANGE

*Duck with orange sauce*

Serves 4

*1 x 1,5 kg duck*
*30 ml melted butter*

**Orange sauce**

*1 orange*
*125 ml fresh unsweetened orange juice*
*15 ml sugar*
*15 ml white-wine vinegar*
*10 ml fresh lemon juice*
*500 ml water*
*1 chicken stock cube*
*15 ml cornflour*
*15 ml water*
*125 ml Grand Marnier or other orange liqueur*

Truss the duck and place it in a roasting pan. Brush with melted butter and roast at 180 °C for about 1 hour, brushing occasionally with pan juices.

To make the sauce, remove the rind from the orange, discarding pith, and cut rind into thin strips. Press out the juice. Combine the rind, orange juice, sugar, vinegar, lemon juice, water and crumbled stock cube in a saucepan. Stir over high heat until the mixture boils, then boil rapidly, uncovered, without stirring, until the mixture is reduced by half.

Blend the cornflour with the extra water and stir into the sauce. Stir constantly until the sauce boils and thickens, about 10 minutes. Remove from the stove and stir in the Grand Marnier.

Drain the pan juices from the roasting pan and pour the sauce over the duck.

Return the pan to the oven and roast for a further 20 minutes, basting often with sauce until the duck is well glazed.

Serve with steamed young green beans and glazed baby onions.

# ROAST CHICKEN ROLL

Serves 6

*1 x 1,5 kg chicken, boned*
*butter*
*salt and freshly ground black pepper*
*paprika*

**Pork and liver stuffing**

*liver and heart from chicken*
*3-4 small onions, quartered*
*300 g minced pork*
*60 ml fresh bread crumbs*
*1 sprig parsley*
*salt and freshly ground black pepper*
*paprika*
*5 ml dried mixed herbs*
*60 ml capers*
*30 ml brandy or sherry*

Remove the wings and drumsticks from the chicken. Prise the meat from them, loosening the meat around the joints with a sharp knife and taking care not to damage the skin. Place the chicken flat, skin side up, and cover it with plastic wrap. Beat the meat carefully with your hands to flatten it.

To make the stuffing, rinse and dry the liver and heart. Mince coarsely, or process in a food processor (but do not allow to become smooth).

Mince or process the onions with the meat from the wings and drumsticks. Combine the chicken meat, liver and heart with the pork, bread crumbs, sprig of parsley, salt, pepper and paprika to taste and dried herbs.

Drain the capers well and mix them into the stuffing with the brandy or sherry.

Remove the plastic wrap from the chicken and spread the stuffing evenly over the meat. Roll up into a thick sausage shape and secure with thin string.

Brown the chicken on all sides in 30 ml butter in a flameproof casserole dish. Season with salt, pepper and paprika to taste.

Place the chicken roll in an ovenproof dish with a little butter and roast at 200 °C for 60-70 minutes, turning several times during cooking.

Transfer the chicken roll to a carving dish, cover loosely and allow to rest for 15 minutes before carving.

Deglaze the ovenproof dish with a little chicken stock to make the gravy.

Serve the chicken hot with buttered young vegetables, mashed potatoes and gravy. To serve cold, slice the chicken roll thinly and serve with French bread and a green salad.

*Right: Chinese stir-fried chicken (p. 62)*

# STUFFED POT-ROASTED MUSCOVY DUCK

Serves 4-6

*1 x 1,5 kg Muscovy duck*
*1 clove garlic, halved*
*6 strips pork fat or bacon*
*10 ml salt*
*2 ml freshly ground black pepper*
*1 onion, sliced*
*5 ml finely chopped fresh mixed herbs*
*50 ml butter or margarine*
*250 ml dry red wine*

**Pork mince stuffing**

*500 g pork mince*
*125 ml fresh bread crumbs*
*100 g rindless streaky bacon, chopped*
*10 ml chopped spring onions*
*1 egg*
*2 ml finely chopped fresh mixed herbs*
*5 ml salt*
*2 ml freshly ground black pepper*

First make the stuffing. Combine all ingredients well and set aside.

Trim the duck and rub inside and out with the garlic clove. Fill the cavity with the stuffing and sew up or close the opening with skewers.

Secure the pork fat or bacon to the breast and back of the duck with toothpicks and dust with salt and pepper.

Place the duck in a heavy-based saucepan with the onion and sprinkle with the herbs. Add the butter and braise the duck, turning once or twice, until golden - about 10 minutes.

Add the wine and simmer, covered, for 2-2½ hours.

Serve with rice, gravy and steamed vegetables.

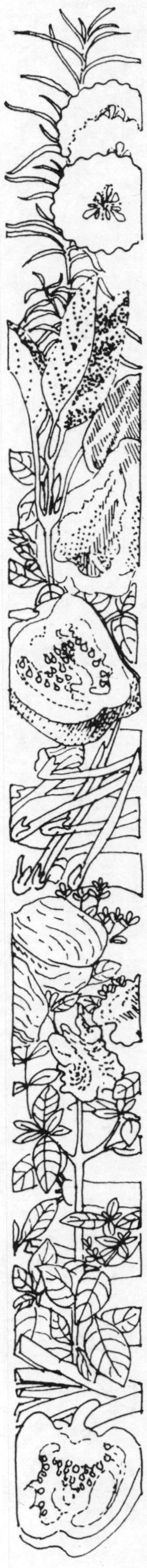

*Left: Spaghetti with green chicken sauce (p. 69)*

## PORT-MARINATED STUFFED TURKEY

Serves 6-8

| | |
|---|---|
| *1 x 2,5 kg* | *turkey* |
| *375 ml* | *dry port* |
| | *salt and freshly ground black pepper* |
| | *stuffing of your choice (see pages 121 to 124)* |
| *100 ml* | *thick cream* |
| *30 ml* | *brown sugar* |
| *50 ml* | *water* |
| *60 g* | *walnuts, coarsely chopped* |

Wipe the turkey with a damp cloth and place it, breast side down, in a dish. Pour the port over and marinate, covered, in a cool place for 8 hours, basting every hour.

Season the turkey inside and out with salt and pepper and stuff it with a stuffing of your choice. Sew up the openings or close them with a skewer.

Roast the turkey as described in the recipe for Oven-roasted turkey (below).

Pour off the pan juices into a saucepan and skim off the fat. Bring to the boil, then stir in the cream and cook, stirring, until smooth.

Place the brown sugar in a separate saucepan, add the water and boil until caramelised – about 15 minutes. Add the nuts and stir for 2 minutes over low heat. Pour over the roast turkey.

Serve the turkey, sliced, with the gravy, parsleyed potatoes and steamed petit pois.

## OVEN-ROASTED TURKEY

Serves 8-10

| | |
|---|---|
| *1 x 2,5 kg* | *turkey* |
| | *stuffing of your choice (see pages 121 to 124)* |
| *15 ml* | *fresh lemon juice* |
| *25 ml* | *salt* |
| *5ml* | *freshly ground black pepper* |
| *30 ml* | *butter or lard* |
| *30 ml* | *cake flour* |

Stuff the body cavity with a stuffing of your choice and sew up the openings or secure them with skewers. Rub the turkey with lemon juice, salt and pepper. Spread it with butter or lard and dredge it in flour.

Pour a little water into the roasting pan, add the turkey and roast, covered, at 160 °C for 2½-3 hours, basting frequently with the pan juices. Uncover the roasting pan for the last 15 minutes to brown the turkey.

Serve with roast potatoes and lightly steamed vegetables.

*Note:*
Use this method to roast any poultry or game bird.

## ROAST STUFFED PIGEONS

Serves 8

*8 pigeons*
*5 ml salt*
*2 ml freshly ground black pepper*
*50 ml lard or margarine*

**Stuffing**

*2 hard-boiled eggs*
*50 ml butter*
*1 slice fresh white bread, crumbled*
*15 ml finely chopped fresh sage*
*5 ml salt*
*2 ml freshly ground black pepper*
*1 egg*

First make the stuffing. Chop the hard-boiled eggs and blend them with the butter, bread crumbs, sage, salt and pepper. Mix in the raw egg to bind.

Lightly season the pigeons inside and out and stuff them. Secure with skewers. Cover the pigeons with lard or margarine and roast at 180 °C for 1¼ hours, basting often.

Serve immediately on a bed of brown rice.

## ROAST GOOSE WITH APPLES AND PRUNES

Serves 6-8

*1 x 3,5 kg goose*
*salt*
*freshly ground black pepper*
*about 100 ml chicken stock**
*3-4 whole cored apples*

**Apple and prune stuffing**
*5-6 apples, peeled, cored and quartered*
*18-20 pitted prunes, soaked until plump and coarsely chopped*

**Gravy**
*5 ml flour mixed to a paste with a little cold water*
*about 100 ml sour cream*
*salt and freshly ground black pepper*

**Garnish**
*125 ml redcurrant or cranberry jelly*

Wipe the inside of the goose with a damp cloth, and season with salt and pepper. Dry the outside with a dry cloth – the skin must be dry otherwise it will not crisp during roasting.

Stuff the goose with apples and prunes and sew up the openings or secure them with skewers. Rub the skin with salt and place the goose on a grid in a roasting pan, back uppermost.

Roast the goose for about 30 minutes at 200 °C, then pour the stock into the roasting pan. Roast the goose for another 30 minutes, then turn it so that the breast is uppermost. Turn the temperature down to 180 °C and continue roasting for another 1¾-2 hours, basting occasionally with the pan juices. Place the whole apples in the roasting pan with the goose for the last 15-20 minutes of the cooking time.

Pour the pan juices into a saucepan and set aside. Turn the temperature up to 250 °C, baste the goose with cold water and leave it in the oven with the door slightly ajar until the skin is brown and crisp. Turn off the oven and leave the goose in the oven with the door open for 5-10 minutes before carving.

Skim the meat juices in the saucepan to remove any fat, then bring it to the boil and thicken it with the flour paste. Boil the gravy for about 5 minutes, then stir in the sour cream and season the gravy with salt and pepper.

Serve the goose with red cabbage, gravy and roast potatoes and the apple sliced and covered with redcurrant or cranberry jelly.

## CHICKEN WITH SPICED FRUIT

Serves 4-6

*1 x 1,5 kg chicken*
*60 ml butter*
*1 small onion, coarsely chopped*
*15 ml grated lemon rind*
*10 ml whole-grain mustard*
*10 ml fresh rosemary, coarsely chopped*
*salt and freshly ground black pepper*
*30 ml sunflower oil*
*2 ml almond essence*

**Sauce**

*50 g dried figs*
*50 g dried apricots*
*50 g dried pears*
*50 g dried peaches*
*50 g dried stoned prunes*
*30 g dried apples*
*375 ml cider or unsweetened apple juice*
*200 ml dry white wine*
*2 pieces root ginger, crushed*
*1 stick cinnamon*
*3 cloves*

Wipe the chicken inside and out with damp absorbent paper. Place half the butter, the onion, lemon rind, mustard and rosemary inside the chicken and season with salt and pepper. Truss the chicken.

Heat the remaining butter and the oil in a roasting pan and add the almond essence. Add the chicken and roast at 180 °C for about 45 minutes, or until tender, basting often with the pan juices.

Meanwhile, prepare the sauce. Place all the dried fruit in a saucepan with the cider or apple juice, wine, crushed ginger, crumbled cinnamon stick and cloves. Bring to the boil over medium heat, then reduce the temperature and simmer, uncovered, until the fruit is soft – about 20 minutes.

When ready to serve, place the chicken on a serving platter, remove the string and keep the chicken warm. Strain the pan drippings and reserve. Using a slotted spoon, remove the fruits from the poaching liquid and arrange them around the chicken.

Add the reserved pan drippings to the liquid in which the fruit was cooked and bring to the boil. Reduce the temperature and simmer for 3 minutes. Remove the cinnamon, cloves and ginger.

Spoon some of the sauce over the chicken and fruits and serve the remaining sauce separately. Couscous or buttered noodles make an excellent accompaniment.

## CHICKEN BALLOTINE

*Vegetable-stuffed chicken drumsticks*

Serves 6

12 *chicken drumsticks, boned*
5 ml *salt*
*freshly ground black pepper to taste*
4 *carrots, scraped*
4 *stalks celery*
4 *baby marrows*
6 *spring onions*
60 ml *butter or margarine*
250 ml *dry white wine*

**Yoghurt and oregano sauce**
250 ml *natural yoghurt*
15 ml *cornflour*
30 ml *brandy*
5 ml *chopped fresh oregano*
2 ml *salt*

Season drumsticks with salt and pepper.

Cut the carrots, celery and baby marrows into thin strips about 10 cm long and halve the spring onions lengthwise. Divide the vegetables and use to stuff the drumsticks. Secure with toothpicks.

Heat the butter or margarine in a large frying pan, add the drumsticks and fry on all sides until golden - 5-7 minutes.

Add the wine and simmer, covered, for 10-15 minutes, or until cooked through. Remove the drumsticks and keep them warm.

To make the sauce, combine the yoghurt and cornflour and stir into the pan juices. Add the brandy, oregano and salt and bring to the boil, then boil, stirring, until thickened - about 3 minutes.

Pour over the chicken and serve on a bed of brown rice.

## CHICKEN KIEV

*Fried chicken breasts stuffed with butter*

Serves 4

4 *chicken breast fillets*
10 ml *salt*
2 ml *freshly ground black pepper*
125 g *butter, quartered*
2 *eggs, beaten*
200 ml *dried bread crumbs*
*olive or sunflower oil*

Season the chicken breasts with salt and pepper and flatten the meat well. Place a lump of butter on each portion of chicken and roll it up firmly, securing it with a toothpick or small skewer.

Coat each breast first in beaten egg and then in bread crumbs.

Heat the oil in a large frying pan and fry the chicken for 7-10 minutes on each side, or until golden brown.

Drain well on absorbent paper and serve immediately with lightly cooked young vegetables.

*Note:*
Finely chopped fresh parsley, tarragon or chives may also be sprinkled on the chicken breasts before rolling them up.

## CHICKEN BREASTS WITH HAM AND CHEESE STUFFING

*This dish, known as Chicken cordon bleu to the French, is a popular item on the menus of restaurants the world over*

Serves 6

| | |
|---|---|
| *6* | *chicken breast fillets* |
| *5 ml* | *salt* |
| | *freshly ground black pepper* |
| *6* | *slices cooked ham* |
| *6* | *slices Gruyère cheese* |
| *1* | *egg, beaten with 30 ml milk* |
| *150 ml* | *dried bread crumbs* |
| *50 ml* | *sunflower oil or butter* |

**Garnish**

*thin lemon slices*
*halved stuffed olives*

Season the chicken with salt and pepper and make a slit in the side of each fillet. Place 1 slice of ham and 1 slice of cheese in each slit and secure them with toothpicks.

Dip the chicken fillets in the beaten egg and then coat them in the bread crumbs. Refrigerate for 1 hour.

Heat the oil or butter in a frying pan and fry the chicken fillets for 10-15 minutes on either side, or until crisp and golden and cooked through.

Garnish with lemon slices and olives and serve with fried chips or rösti*.

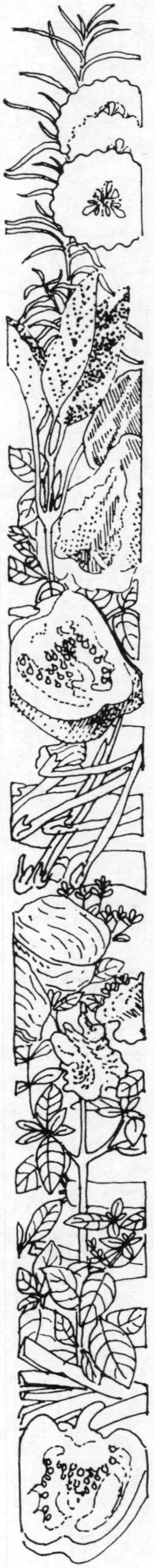

## TURKEY STEAKS WITH GREEN-PEPPERCORN SAUCE

Serves 4-6

*250 ml chicken stock**
*60 ml finely chopped leeks*
*8 boned turkey breast steaks*
*60 ml butter or margarine*
*5 ml salt*
*60 ml dry white vermouth*
*30 ml thick cream*
*30 ml green peppercorns*
*5 ml fresh lemon juice*

Bring the chicken stock to the boil and boil rapidly, uncovered, with the leeks until reduced to 100 ml.

Flatten the turkey steaks and fry on both sides in 15 ml heated butter or margarine until crisp and golden, about 10 minutes. Remove the steaks from the pan, sprinkle with salt and keep warm.

Strain the stock and add it to the pan juices. Bring to the boil, then add the vermouth and boil rapidly, uncovered, until reduced to about 50 ml.

Add the cream, peppercorns and lemon juice and whisk in the remaining butter. Pour over the steaks and serve immediately with rice.

*Note:*
Chicken steaks may be used instead of turkey.

## DEEP-FRIED CHICKEN WITH WALNUTS

Serves 6

*6 chicken breasts, skinned, boned and diced*
*5 ml salt*
*15 ml cornflour*
*1 egg white*
*125 g blanched walnuts*
*sunflower oil for deep-frying*
*1 green pepper, seeded and diced*
*1 red sweet pepper, seeded and diced*
*15 ml sunflower oil*
*30 ml sugar*
*30 ml soy sauce*
*15 ml dry white wine*

Mix the chicken with the salt, cornflour and egg white, tossing well.

Put the walnuts in a strainer and deep-fry them in oil heated to 150 °C until light brown and crisp. Drain well and set aside. (Take care not to brown them too much, as they burn easily.)

Sauté the green and red peppers in 15 ml oil in a frying pan for 1 minute, drain well and set aside. Add the sugar to the frying pan.

Deep-fry the chicken in oil heated to 190 °C for 1 minute. Drain well and add, with the soy sauce, wine, peppers and walnuts, to the frying pan.

Heat through, tossing constantly, and serve with a spinach salad.

*Note:*
The walnuts and chicken deep-fry best in an automatic deep-frier, which has a built-in temperature gauge. If you don't have one, use a thermometer to test the temperature, or fry a cube of bread in hot oil – the oil has reached the right temperature if the bread browns in 45-55 seconds.

## GINGERED CHICKEN FRY

Serves 4

| | |
|---|---|
| 8 | *chicken portions* |
| | *salt and freshly ground black pepper* |
| 1 | *clove garlic, crushed* |
| 5 ml | *ground ginger* |
| 30 ml | *butter* |
| 30 ml | *sunflower oil* |
| 60 g | *fresh root ginger, peeled and thinly sliced* |
| 1 | *green pepper, seeded and cut into strips* |
| 3-4 | *ripe tomatoes, skinned and quartered* |
| 100-200 ml | *chicken stock** |
| 30 ml | *chopped raw cashew nuts* |

Rub the chicken portions with salt, pepper and garlic, and sprinkle with ground ginger.

Heat the butter and oil in a large frying pan and sauté the chicken on all sides until golden – about 10 minutes.

Add the ginger and vegetables to the frying pan, pour over 100 ml stock and simmer, covered, over medium heat until the meat is tender – about 30 minutes. Turn the chicken often, adding more stock as needed.

About 5 minutes before the end of the cooking time, add the cashew nuts and cook, uncovered.

Serve straight from the pan with noodles tossed in garlic butter and a green salad.

## FRUITY CHICKEN WITH SAUSAGES

Serves 4

*1 x 1,5 kg chicken, cut into portions*
*5 ml salt*
*freshly ground black pepper*
*30 ml sunflower oil*
*1 onion, chopped*
*1 carrot, chopped*
*1 stalk celery, chopped*
*30 ml cake flour*
*200 ml dry white wine*
*250 ml chicken stock**
*200 g cocktail sausages*
*15 ml margarine*
*100 ml chopped dried apricots*
*50 ml raisins*
*50 ml chopped stoned prunes*

Season the chicken with the salt and pepper. Heat the oil in a heavy-based saucepan and fry the chicken on either side for 5 minutes, or until golden, then remove from the pan and keep warm.

Add the onion, carrot and celery to the pan and sauté until softened, about 3 minutes.

Mix the flour and wine to a paste and stir into the vegetables. Add the stock and boil for 7-10 minutes, or until the sauce thickens, stirring constantly. Add the chicken to the sauce.

In a frying pan, fry the sausages in the heated margarine, turning often, for 5-7 minutes, or until golden all over and cooked through. Add the dried fruit and stir until the fruit is heated through, about 5 minutes.

Add the sausages and fruit to the chicken and heat through, about 2 minutes.

Serve immediately with brown or yellow rice.

## ROSEMARY-FRIED CHICKEN WITH MEALIE FRITTERS

*Our variation of the American favourite, chicken Maryland*

Serves 8

*300 ml cake flour*
*10 ml paprika*
*7 ml salt*
*7 ml dried thyme*
*5 ml crumbled dried rosemary, or chopped fresh rosemary*
*1 egg, beaten*
*50 ml milk*
*30 ml fresh lemon juice*
*16 chicken portions*
*sunflower oil*

**Mealie fritters**
*250 ml coarse mealie meal*
*250 ml boiling water*
*65 ml cake flour*
*15 ml baking powder*
*5 ml salt*
*5 ml sugar*
*1 egg, beaten*
*150 ml milk*

First prepare the chicken. Combine the flour, paprika, salt and herbs. Mix the egg, milk and lemon juice well. Coat the chicken portions in the flour mixture, dip them in the egg mixture and then coat the portions in the flour mixture again. Refrigerate for at least 30 minutes.

Heat the oil, 1 cm deep, in a large frying pan and fry the chicken until golden brown, about 5 minutes on either side. Reduce the temperature, cover the frying pan and continue to cook for 25 minutes.

Meanwhile, make the mealie fritters. Mix the mealie meal with water in a bowl. Cool for 10 minutes, then add the remaining ingredients and beat until smooth. The batter should be thin.

Drop tablespoonfuls of batter onto a hot greased pan or griddle. Cook until brown underneath, about 4 minutes, then turn and brown them on the other side, about 3 minutes.

Drain the chicken well on absorbent paper and serve at once with the mealie fritters and a salad.

## PARMESAN CHICKEN

Serves 6

*6 chicken breast fillets*
*100 ml cake flour*
*5 ml salt*
*1 ml freshly ground white pepper*
*2 eggs*
*50 g grated Parmesan cheese*
*30 ml milk*
*30 ml butter or margarine*
*50 ml sunflower oil*

**Garnish**
*12 thin lemon slices*

Flatten the chicken breasts lightly with a meat mallet or your hand.

Combine the flour, salt and pepper in a small bowl.

Beat the eggs in a bowl and mix in the Parmesan cheese and milk.

Heat the butter or margarine and oil in a large frying pan over medium heat.

Dip the chicken breasts first in flour and then in the cheese batter, turning to coat both sides well.

Fry the chicken in the heated butter and oil until golden, about 5 minutes, then turn the pieces over. Reduce the temperature to low and continue to cook until the chicken is cooked through, about 10 minutes.

Serve immediately, topped with lemon slices and accompanied by stir-fried vegetables and minted new potatoes.

## CHICKEN IN BEER WITH PEACHES

Serves 4

*1 x 1-1,5 kg chicken*
*50 ml butter or margarine*
*salt and freshly ground black pepper*
*15 ml chopped spring onions*
*30 ml vodka*
*100 ml thick cream*
*250 g baby mealies, diced*
*4 yellow cling peaches, peeled, stoned and sliced*
*500 ml draught beer or ale*
*cayenne pepper*

Truss the chicken. Heat 30 ml of the butter or margarine in a flameproof casserole dish and turn the chicken in it over high heat until golden, about 10 minutes.

Season to taste with salt and pepper, cover and cook for 30 minutes over low heat. Remove the chicken to a heated dish, cover and keep it warm.

Sauté the spring onions in the casserole until softened, 2-3 minutes.

Replace the chicken, pour the vodka over and flame it.

Add 15 ml butter, 30 ml of the cream, the mealies and peaches. Pour in the beer and season with salt, pepper and a little cayenne pepper.

Simmer, covered, for 15 minutes, or until the chicken is tender.

Remove the chicken to a chopping board and cut into 4 pieces. Place on a heated serving dish and keep warm.

Drain the peaches and mealies and keep warm with the chicken.

Pour the remaining cream into the casserole dish containing the cooking liquor and boil vigorously until thickened, about 10 minutes.

Adjust the seasoning if necessary, and take the casserole off the stove. Add the remaining butter or margarine and let it melt in the sauce, off the heat.

Pour the sauce over the chicken and serve immediately with crushed wheat.

*Note:*
Canned peaches and mealies may be used if fresh ones are not available.

## COQ AU VIN

*Chicken in red wine*

Serves 4-6

*15 ml butter or margarine*
*20 ml sunflower oil*
*1 x 1,5 kg chicken, cleaned and giblets removed*
*4 rashers rindless bacon, chopped*
*16-18 pickling onions*
*2 cloves garlic, crushed*
*375 ml dry red wine*
*250 ml chicken stock**
*15 ml tomato paste*
*250 g button mushrooms*
*30 ml brandy*
*15 ml cake flour blended into 30 ml water*

Heat the butter or margarine and oil in a large deep saucepan and brown the chicken, about 5 minutes on either side.

Remove it from the pan and pour off most of the fat, then add the bacon, onions and garlic to the pan and sauté them until the onions are transparent - about 5 minutes.

Add the wine, stock and tomato paste and return the chicken to the pan. Simmer, covered, for 35-40 minutes, then add the mushrooms and simmer for a further 10 minutes. Remove the chicken and vegetables to a serving platter.

Add the brandy and flour mixture to the saucepan and boil, stirring constantly, until thickened.

Pour the sauce over the chicken and serve with crusty French bread.

## CHICKEN WITH FRESH HERBS AND CREAM

Serves 4-6

*60 ml butter or margarine*
*1 x 1,5 kg chicken, cut into portions*
*5 spring onions*
*100 ml Madeira wine or sweet sherry*
*100 ml brandy*
*30 ml tomato purée*
*150 ml thick cream*
*salt and freshly ground white pepper*
*ground ginger*
*a pinch each finely chopped fresh tarragon, basil and savory, or other herbs of your choice*

Melt the butter or margarine in a flameproof casserole dish, add the chicken portions and brown them quickly – about 5 minutes on either side.

Add the spring onions, then stir in the Madeira wine or sherry, brandy, tomato purée and cream. Season to taste with salt and pepper, ginger and herbs.

Cover the casserole dish and bring to the boil, then bake in the oven at 180 °C for 20 minutes.

Serve with young baby mealies and steamed young green beans.

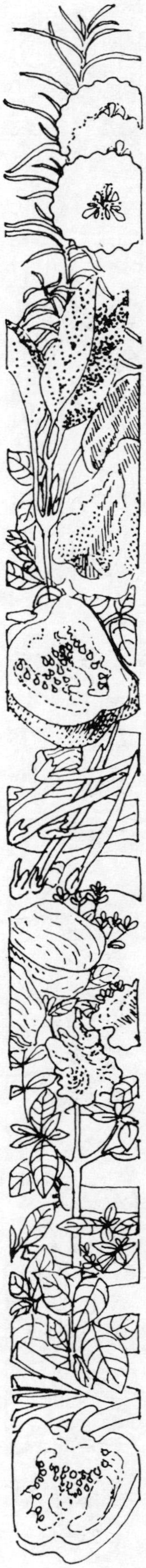

## BRAISED CHICKEN THIGHS WITH MUSHROOMS

Serves 4

8 *chicken thighs*
*seasoned flour*
20 ml *sunflower oil*
½ *onion, chopped*
200 g *brown mushrooms, sliced*
2 *cloves garlic, crushed*
50 ml *tomato paste*
2 ml *crumbled dried rosemary*
2 ml *crumbled dried thyme*
250 ml *dry white wine*
200 ml *chicken stock**

**Garnish**
*chopped fresh parsley or watercress sprigs*

Dry the chicken thighs and dredge them in the flour, shaking off the excess.

Heat the oil in a flameproof casserole dish over medium heat until hot but not smoking and brown the chicken for 5 minutes on either side. Transfer the thighs to a plate.

Add the onion, mushrooms, garlic, tomato paste, rosemary and thyme to the casserole dish and cook over medium heat, stirring, for 1 minute.

Return the chicken, skin side up, to the casserole, add the wine and chicken stock and bring to the boil.

Transfer to the oven and braise at 200 °C for 20 minutes, or until the chicken is tender.

Adjust the seasoning if necessary and serve on a bed of buttered noodles, garnished with the parsley or watercress sprigs.

## CHICKEN À LA KING

*Chicken with mushrooms and green pepper*

Serves 4-6

*50 ml butter or margarine*
*1 green pepper, seeded and sliced*
*125 g mushrooms, sliced*
*15 ml cake flour*
*275 ml chicken stock**
*150 ml milk*
*1 x 1,5 kg chicken, cooked, boned and diced*
*5 ml salt*
*2 ml freshly ground white pepper*
*2 egg yolks*
*150 ml cream*

Melt the butter or margarine in a large frying pan and sauté the green pepper in it until soft, about 5 minutes. Add the mushrooms and sauté until tender, about 5 minutes.

Remove the frying pan from the stove, blend in the flour and then gradually stir in first the chicken stock and then the milk. Return the frying pan to the stove and bring the mixture to the boil, stirring continuously until thickened – 5-7 minutes. Add the chicken and the salt and pepper and heat through, about 10 minutes.

Beat the egg yolks and cream together lightly in a bowl, stir some of the hot sauce into it and heat through, but do not boil.

Serve immediately with rice or noodles and tomato salad.

## GARLIC CHICKEN IN YOGHURT

Serves 4

*1 x 1,5 kg chicken, skinned and cut into portions*
*75 ml chopped fresh parsley or coriander leaves*

**Marinade**
*350 ml natural yoghurt*
*1 large red sweet pepper, seeded and very finely chopped*
*1 x 5 cm piece root ginger, grated*
*2-3 green chillies, crushed*
*16 large cloves garlic, crushed*
*5 ml paprika*
*5 ml salt*

*Right: Lemon chicken in a clay pot (p. 78)*

Prick the chicken pieces all over with a very sharp fork.

Combine all the marinade ingredients and marinate the chicken pieces in it, in the refrigerator, for 10-12 hours, turning from time to time.

Heat a heavy-based saucepan until very hot. Add the chicken and marinade, then stir in the parsley or coriander leaves and cover. Cook on high heat for 5 minutes, then reduce temperature to medium and continue cooking until almost all the marinade has evaporated, leaving 15-30 ml in the bottom. Do not allow to brown. Turn the chicken occasionally to coat it evenly.

Serve hot, with the scrapings from the saucepan, accompanied by rice.

## PORTUGUESE CHICKEN

Serves 6

*2 x 1,5 kg chickens, cut into portions*
*10 ml salt*
*2 ml freshly ground black pepper*
*30 ml sunflower oil*

**Tomato mushroom sauce**
*1 medium onion, sliced*
*5 ml crushed garlic*
*125 g mushrooms, sliced*
*100 ml dry white wine*
*3 ripe tomatoes, skinned and coarsely chopped*
*5 ml salt*
*1 ml freshly ground black pepper*
*15 ml chopped fresh thyme*
*125 g black olives*

Season the chicken portions with salt and pepper. Heat the oil in a frying pan and brown the chicken for 5 minutes on either side. Remove from the pan and keep warm and covered.

To make the sauce, add the onion, garlic and mushrooms to the pan and sauté until soft but not brown - about 5 minutes.

Add the white wine and boil vigorously, uncovered, until reduced to a quarter of the original volume. Add the tomatoes and bring to the boil. Boil vigorously to reduce slightly, then season with salt and pepper.

Place the chicken portions in the sauce and simmer, covered, over low heat until cooked through - about 30 minutes. Add the thyme and olives.

Serve the chicken topped with the sauce on a bed of buttered noodles.

*Note:*
The chicken may also be baked in the sauce in the oven at 160 °C for 30-45 minutes.

*Left: Paella over the coals (p. 108)*

## CHICKEN CASSEROLE WITH HERB DUMPLINGS

Serves 6

6 *chicken portions*
125 ml *cake flour seasoned with 5 ml salt and 5 ml pepper*
65 ml *butter or margarine*
8 *small onions*
3 *stalks celery, finely chopped*
500 ml *water*
1 *lemon, thinly sliced*
1 *bay leaf*
180 g *carrots, thinly sliced*
125 g *green peas*
150 ml *sour cream or natural yoghurt*

**Herb dumplings**
125 g *self-raising flour*
2 ml *salt*
60 g *suet*
5 ml *dried mixed herbs*
*water to mix*

**Garnish**
*finely chopped fresh parsley*

Coat the chicken pieces in seasoned flour, reserving excess flour.

Melt the butter or margarine in a large frying pan and fry the chicken for 3-4 minutes on either side, or until golden brown.

Place the chicken portions in a large deep casserole dish. Add the onions and celery to the frying pan and sauté until the onions are transparent, about 5 minutes. Add to the chicken in the casserole dish.

Stir the remaining flour into the butter or margarine in the frying pan, then gradually stir in the water. Bring to the boil and continue boiling until thickened, about 5 minutes, stirring constantly.

Pour the sauce over the chicken and add the lemon and bay leaf. Bake at 180 °C, covered, for 1 hour, or until the chicken is tender.

Meanwhile, boil the carrots and peas in salted water until barely tender. Drain and add them to the casserole dish.

While the casserole is cooking, make the dumplings. Sift the flour and salt into a bowl. Cut in the suet with a knife, blend in the herbs and stir in enough water to bind. The dough should be soft enough to form into balls.

Divide into 8-10 portions and with floured hands, roll them into balls. About 30 minutes before the end of the cooking time, add the dumplings to the casserole, making sure that the liquid is boiling steadily.

Remove the lemon slices and bay leaf from the casserole and pour the sour cream or yoghurt over.

Garnish with chopped parsley and serve with crushed wheat.

## CHICKEN-STUFFED CABBAGE ROLLS

Serves 4

4 *large or 8 smaller cabbage leaves*
250 ml *chicken stock**

**Stuffing**
1 *medium onion, chopped*
*sunflower oil*
250 g *chicken, minced*
1 *Granny Smith apple, cored, peeled and finely chopped*
45 ml *fruit chutney*
1 *egg*
*salt and freshly ground black pepper*

Remove the coarse ribs from the cabbage and blanch the leaves for 5-7 minutes in lightly salted water to cover. Drain well.

To make the stuffing, sauté the onion in a little oil until transparent – about 5 minutes.

Add the chicken, apple and chutney and stir over medium heat until well combined and the chicken is cooked, about 10 minutes. Remove from the stove and set aside to cool.

Add the egg and salt and pepper to taste and mix well.

Form the mixture into 4 large or 8 small patties and place 1 in each cabbage leaf. Roll up to enclose the stuffing completely and secure with toothpicks.

Place cabbage rolls, seam side down, close together in a saucepan and pour the stock over them.

Simmer over low heat, partially covered, until the cabbage is tender – about 15 minutes.

Serve immediately with fried rice and steamed butternut squash.

# CHICKEN BREYANI

Serves 4-6

*200 ml sunflower oil*
*6 medium onions, finely chopped*
*6 medium potatoes, cubed*
*250 g uncooked rice*
*125 ml red lentils*
*5 ml salt*
*1 x 1,5 kg chicken, cut into portions*
*7 ml ground coriander*
*7 ml turmeric*
*2 ml saffron*
*1 stick cinnamon*
*2 bay leaves*
*5 cardamom seeds*
*2 ml cumin seeds*
*65 ml fresh lemon juice*
*3 ripe tomatoes, skinned and finely chopped*
*7 ml finely chopped chillies*
*5 ml salt*
*a few black peppercorns*
*4 cloves garlic*
*250 ml natural yoghurt*
*1 piece fresh root ginger, crushed*
*4 hard-boiled eggs, thinly sliced*

Heat 15 ml oil in a frying pan and brown the onions, about 5 minutes. Remove the onions and set them aside. Add the potatoes and sauté until lightly browned, 5-7 minutes. Remove the potatoes, drain them on absorbent paper and set aside.

Boil the rice in boiling salted water until almost tender but still firm, about 10 minutes. Drain, and add more salt, if necessary.

In another saucepan, boil the lentils for 10 minutes in salted water to cover, then drain well.

Combine the rice, ¼ of the fried onions and the lentils in a bowl. Rub a little salt into the chicken pieces and place them in a large mixing bowl. Combine the herbs, spices, lemon juice, tomatoes, chillies, 5 ml salt, peppercorns, the rest of the fried onions, the garlic, yoghurt and ginger in a bowl. Pour over the chicken and leave for 5 minutes.

Invert a heavy dinner plate in a large saucepan so that it covers the bottom completely. Arrange the ingredients on top of the plate in the following order: half the browned potatoes, sprinkled with 5 ml oil; chicken pieces and sauce; hard-boiled eggs; rice and lentil mixture; the rest of the oil sprinkled over the contents of the saucepan.

Cover with a tight-fitting lid and simmer over low heat for 1½-2 hours.

Serve at once with apple salad, banana slices dipped in lemon juice, tomato wedges, desiccated coconut and chutney.

## CHICKEN CURRY

Serves 6

*1 x 1,5 kg chicken, cut into portions*
*2 medium onions, finely chopped*
*30 ml sunflower oil*
*30 ml butter*
*500 ml chicken stock**
*5 ml salt*
*30 ml curry powder*
*10 ml turmeric*
*10 ml ground coriander*
*125 ml seedless raisins*
*125 ml desiccated coconut*
*5 ml sugar*
*2 ml ground cinnamon*
*125 ml cream*
*15 ml cornflour*
*15 ml brandy*

Fry the chicken portions and onions in the heated oil and butter in a large saucepan until golden brown, 5-7 minutes, turning the chicken once.

Combine the stock, salt, curry powder, turmeric, coriander, raisins, coconut, sugar and cinnamon in a bowl.

Transfer the chicken, onion and pan juices to a large casserole dish and pour the curry mixture over. Bake, covered, at 160 °C until tender – about 1 hour.

Mix the cream, cornflour and brandy in a small saucepan and simmer over medium heat, stirring, until thickened – 5-7 minutes.

Pour the sauce over the chicken and serve with boiled rice and fruit chutney.

*Note:*
The curry may also be cooked in the saucepan on the stove. Simmer over low heat for 1 hour.

## CHICKEN WITH SEAFOOD

Serves 6

*8-12 chicken portions*
*50 ml cake flour*
*30 ml butter or margarine*
*30 ml sunflower oil*
*3 large tomatoes, skinned and chopped*
*250 g button mushrooms, sliced*
*1 onion, chopped*
*2 cloves garlic, crushed*
*125 ml water*
*1 chicken stock cube, crumbled*
*250 ml dry white wine*
*30 ml tomato paste*
*15 ml chopped fresh basil*
*30 ml brandy*
*500 g cooked, shelled prawns*

Toss the chicken portions in flour. Heat the butter or margarine and oil in a saucepan and fry the chicken until golden, about 5 minutes on either side.

Drain the fat from the saucepan and add the tomatoes, mushrooms, onion, garlic, water, crumbled stock cube, wine, tomato paste and basil. Simmer, covered, for 30 minutes or until tender.

Stir in the brandy and turn out into a serving dish. Heat the prawns and place on top of the chicken.

Serve with buttered noodles.

*Note:*
To cut down on cooking time, microwave the chicken on MEDIUM-HIGH for 15 minutes after adding all the ingredients except the brandy and prawns. Stir in the brandy, microwave the prawns on HIGH for 3-4 minutes and add to the dish.

## CHICKEN WITH BRINJALS

Serves 4-6

*1 x 2 kg chicken, cut into portions*
*60 ml olive oil*
*5 cloves garlic, crushed*
*salt and freshly ground black pepper*
*2 ml ground saffron*
*3 medium brinjals, peeled and cubed*
*150 ml water*

Sauté the chicken in the heated olive oil, with the garlic, until browned – about 5 minutes on either side.

Season with salt, pepper and saffron, and add the cubed brinjals. Add the water, then simmer, covered, over low heat for 20 minutes, or until the chicken is tender.

Serve on a bed of brown rice or couscous.

## BRAISED CHICKEN ROLLS

Serves 6

*6 boned chicken breasts*
*5 ml salt*
*30 ml butter*
*325 ml chicken stock**
*125 ml dry white wine*
*10 ml cornflour*

**Stuffing**
*325 ml fresh bread crumbs*
*4 rashers rindless bacon, finely chopped*
*1 egg, lightly beaten and mixed with 50 ml milk*
*5 ml chopped fresh savory or oregano*
*15 ml chopped fresh parsley*
*15 ml butter*

**Garnish**
*chopped fresh parsley*

Flatten the chicken breasts with a meat mallet and season with salt.

Combine all the stuffing ingredients well and divide between the breasts. Roll up the chicken and secure with toothpicks.

Melt the butter in a deep frying pan and brown the chicken rolls, about 5 minutes on either side.

Add the stock and wine and simmer, covered, for 30 minutes. Remove the rolls from the stock and keep them warm.

Thicken the stock slightly with cornflour and pour over the rolls.

Sprinkle with chopped fresh parsley and serve with mashed potatoes.

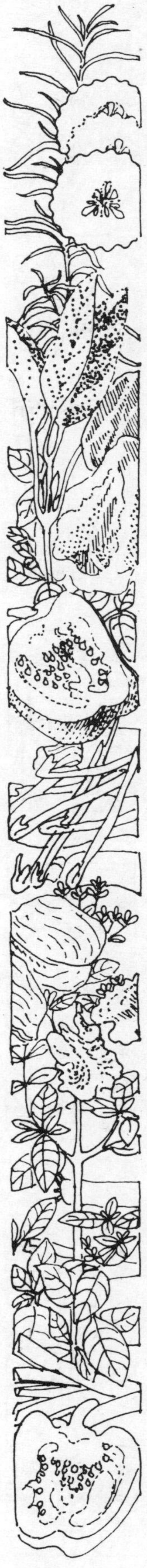

## SWISS TURKEY

Serves 4-6

*400-600 g skinned turkey breast meat*
*salt and freshly ground black pepper*
*butter*
*500 g young spinach or Swiss chard*
*grated nutmeg*
*4 thick slices cooked ham*
*4 slices Gruyère or mozzarella cheese*
*200 ml chicken stock**
*150 ml dry white wine*

Beat turkey breasts flat and quarter them. Rub a little salt and pepper into each and brown them slightly on both sides in heated butter, about 5-8 minutes.

Rinse the spinach or chard well, remove stalks and shred the leaves coarsely. Sauté in heated butter for 5 minutes and season with salt, pepper and nutmeg.

Halve the ham slices and place half of them in the base of a greased ovenproof dish. Spread the spinach on top of the ham, then top with the turkey meat and finally the remaining ham. Cover with the cheese slices and pour over the stock and wine.

Bake at 180 °C for 30 minutes, or until the cheese is golden and the dish is heated through.

Serve immediately with French bread, chicory and watercress salad.

*Note:*
This dish is ideal for using up leftover roast turkey. Reduce the cooking time to 15 minutes and do not flatten the meat.

## TURKEY FRICADELS WITH SOUR-CREAM SAUCE

Serves 4-6

*500 g minced turkey*
*1 egg*
*125 ml cream*
*1 Granny Smith apple, peeled, cored and grated*
*2 ml grated nutmeg*
*1 egg, lightly beaten*
*65 ml milk*
*375 ml dried bread crumbs*
*60 ml butter*
*30 ml sunflower oil*

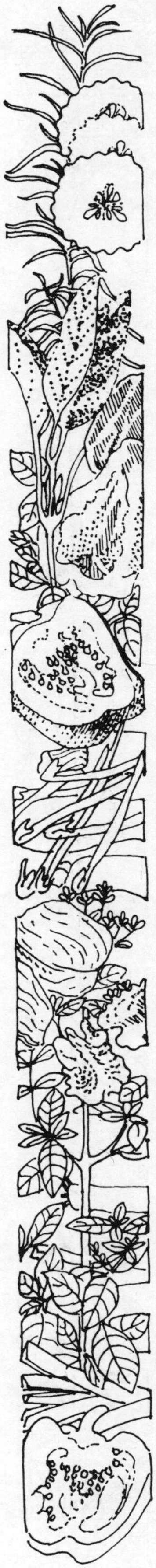

**Sour-cream sauce**

*30 ml butter*
*1 onion, finely chopped*
*30 ml white vinegar*
*30 ml fresh lemon juice*
*10 ml cake flour*
*300 ml thick cream*

Combine the turkey, egg and cream in a large bowl. Mix in the grated apple and nutmeg and form into 4-6 fricadels.

Combine the egg and milk and dip the fricadels in it, then coat them in bread crumbs, pressing the crumbs in firmly.

Heat the butter and oil in a frying pan and fry the fricadels over medium heat for 5 minutes on each side. Drain the fricadels on absorbent paper and place them on a baking sheet. Bake for 15 minutes at 180 °C, or until golden.

Meanwhile, make the sour-cream sauce. Heat the butter in a small saucepan and sauté the onion until soft, about 5 minutes. Add the vinegar and lemon juice and cook for 1 minute. Blend the flour and cream and stir it into the sauce. Cook, stirring, until the sauce boils and thickens, about 3 minutes.

Serve the fricadels hot, with the sauce poured over them, and accompanied by buttered spinach ribbon noodles and baked stuffed tomatoes.

*Note:*
Any other uncooked poultry can be used to make these fricadels.

## PHEASANT CASSEROLE

Serves 4

*100 ml butter or margarine*
*1 x 1,5 kg pheasant, cut into portions*
*1 large onion, quartered*
*250 ml thick cream*
*juice of ½ lemon*
*5 ml paprika*
*10 ml salt*
*5 ml freshly ground black pepper*
*5 ml dried mixed herbs*

Melt the butter or margarine in a large flameproof casserole dish, add the pheasant and onion and braise, covered, over medium heat for 50 minutes.

Stir in the cream, lemon juice and paprika, season with salt, pepper and herbs and simmer, covered, over very low heat until the pheasant is tender – about 15 minutes.

Serve immediately with brown rice.

*Note:*
Guinea fowl may be cooked in the same way.

## CHICKEN RATATOUILLE

Serves 4-6

*75 ml olive oil*
*1 x 1,5 kg chicken, cut into portions*
*2 large onions, sliced*
*3 cloves of garlic*
*300 g green peppers, seeded and sliced into strips*
*500 g ripe tomatoes, skinned, seeded and chopped*
*5 ml fresh thyme, chopped*
*salt and freshly ground black pepper*
*cake flour*
*300 g baby marrows, trimmed and cubed*
*300 g brinjals, trimmed, cubed and soaked in salted water*

**Garnish**

*chopped fresh thyme or basil*

Heat 30 ml oil in a frying pan and sauté the chicken pieces until golden brown, about 5 minutes on either side.

Meanwhile, heat 30 ml oil in a flameproof casserole dish or large saucepan and sauté the onions, 2 cloves of garlic and green peppers until softened, about 5 minutes.

Add the tomatoes and thyme and continue cooking until the vegetables are almost tender, 5-7 minutes.

Add the chicken, season to taste and simmer, covered for 45 minutes.

Meanwhile, lightly flour the baby marrows and drained brinjals and sauté them until golden in the oil remaining in the frying pan, about 5 minutes, adding more oil if necessary.

Arrange the chicken and vegetables in a heated serving dish and garnish with the baby marrows and brinjals.

Chop the remaining garlic clove finely and sprinkle, with the thyme or basil, over the chicken and vegetables.

Serve immediately, with crusty French bread.

## BRAISED DUCK WITH YELLOW PLUMS

Serves 6

*100 ml butter or margarine*
*1 x 2 kg duck, cleaned and trussed*
*300 ml white muscadel*
*5 ml salt*
*5 ml freshly ground black pepper*
*1 kg yellow plums, halved and stoned*

Melt half the butter or margarine in a deep casserole dish and sauté the duck over low heat, turning occasionally, until evenly browned – about 15 minutes.

Add the muscadel, salt and pepper and braise the duck in the oven at 200 °C until tender, about 45 minutes, basting often.

Skim excess fat from the surface of the cooking juices and add more salt and pepper if necessary.

Meanwhile, sauté the plums in the remaining butter or margarine in a frying pan until just cooked, about 5 minutes on either side.

Place the duck on a serving dish and decorate with the plum halves. Strain the pan juices and pour over the duck.

Serve at once with buttered noodles.

## QUAIL CASSEROLE WITH BACON AND MUSHROOMS

Serves 6

| | |
|---|---|
| 4-6 | *quail* |
| 5 ml | *salt* |
| 2 ml | *freshly ground white pepper* |
| 60 ml | *butter or margarine* |
| 1 | *leek, thinly sliced* |
| 1 | *green pepper, seeded and sliced* |
| 125 g | *mushrooms, chopped* |
| 6 | *rashers rindless bacon, chopped* |
| 5 ml | *chopped fresh oregano* |
| 125 ml | *dry white wine or chicken stock** |
| 30 ml | *fresh lemon juice* |

Split the quail in half and flatten them well. Season with salt and pepper.

Heat 45 ml butter or margarine in a frying pan and sauté the leek, green pepper, mushrooms, bacon and oregano until softened – 5-10 minutes.

Spoon the mixture into a casserole dish and place the quail on top. Pour over the wine or stock and lemon juice.

Melt the remaining butter and use to brush the quail. Bake, covered, at 180 °C for 45 minutes. Remove the lid and bake for a further 15 minutes.

Serve with buttered noodles.

## PAELLA OVER THE COALS

Serves 6

*6 chicken portions, seasoned with salt and pepper*
*15 ml olive oil*
*1 onion, sliced*
*1 tomato, skinned and chopped*
*1 small green pepper, seeded and chopped*
*2 cloves garlic, crushed*
*10 ml chopped fresh mixed herbs*
*5 ml salt*
*125 g uncooked rice*
*1 chicken stock cube dissolved in 600 ml water*
*1 pinch saffron or 5 ml turmeric*
*2 small uncooked crayfish, split in half*
*1 x 110 g can smoked oysters*
*12 unshelled mussels, scrubbed and debearded*
*12 black olives*
*2 smoked Russian sausages, sliced*

Braai the chicken over the coals, turning often, until just done - 20-25 minutes. Set aside and keep warm.

Heat the oil in a paella pan or large cast-iron pot and sauté the onion, tomato, green pepper, garlic, herbs and salt in it for 5 minutes.

Add the rice and fry for 2 minutes. Add the water in which the stock cube has been dissolved and the saffron or turmeric.

Arrange the seafood, olives and sausage on top of the rice mixture and cook slowly over the coals until the rice is cooked and fluffy - 20-25 minutes. Place the chicken on top and serve with a green salad.

*Notes:*

Paella may also be cooked in the oven or on top of the stove.

Any selection of seafoods may be used - sliced calamari and prawns are particularly good.

If cooked seafood is used, add it about 10 minutes before the end of the cooking time.

## TANDOORI CHICKEN

*An Indian dish, traditionally cooked in a cylindrical clay oven heated to a high temperature by a wood and coal fire*

Serves 6

*3 x 1 kg chickens*
*7 ml salt*
*50 ml fresh lemon juice*
*50 ml melted butter or sunflower oil*

**Tandoori marinade**

*500 ml natural yoghurt*
*5 cloves garlic, crushed*
*5 ml cayenne pepper*
*5 ml ground ginger*
*10 ml ground cumin*
*5 ml ground coriander*
*5 ml finely chopped fresh mixed herbs*
*5 ml red food colouring*

**Garnish**

*lemon wedges*

Skin and halve the chickens. Make incisions in the meat and rub well with salt and lemon juice.

Combine all the marinade ingredients and marinate the chicken halves for at least 5 hours, but preferably overnight.

Remove the chickens from the marinade and braai them over very hot coals or grill them in the oven until golden, basting frequently with the butter or oil.

Serve immediately on rice, garnished with lemon wedges, accompanied by a crisp green salad.

*Note:*
The marinade may be thickened with a little cornflour and served, heated, with the chicken.

## CITRUS-MARINATED GRILLED BUTTERFLY CHICKEN

Serves 4

*2 x 1 kg chickens, split down the back and flattened*

**Citrus marinade**

*100 ml dry white wine*
*100 ml white-wine vinegar*
*100 ml fresh orange juice*
*50 ml sunflower oil*
*15 ml dried mixed herbs*
*10 ml salt*
*2 ml cayenne pepper*
*1 large onion, sliced*

Combine the marinade ingredients and marinate the chickens for 2-3 hours, turning frequently.

Remove the chickens from the marinade and grill them for 30 minutes over hot coals, turning and basting often with the marinade.

Boil and strain the marinade and serve it as a sauce with the chicken, accompanied by steamed mealie bread.

## WHOLE CHICKEN ON THE SPIT

Serves 4

*1 x 1,5 kg chicken*

**Honey and garlic marinade**

*1 onion, finely chopped*
*2 cloves garlic, crushed*
*50 ml sunflower oil*
*15 ml honey*
*50 ml tomato sauce*
*30 ml Worcester sauce*
*15 ml fresh lemon juice*
*5 ml paprika*
*5 ml Aromat or Fondor*

Tie the wing tips over the breast of the chicken and tie the neck skin to the back with a skewer or string. Push the spit through the chicken from the tail end. Cross the drumsticks and tie to the tail to anchor the chicken to the spit.

Combine all the marinade ingredients in a jug.

Grill the chicken over the coals for about 1 hour, turning and basting often with the marinade.

Serve with baked potatoes and a mixed salad.

*Notes:*

The chicken may also be stuffed with a stuffing of your choice (see pages 121 to 124); allow 15-20 minutes extra cooking time.

To shorten the cooking time, the chicken can be parboiled for about 10 minutes and grilled for the last 20-30 minutes.

## GREEK-STYLE GRILLED CHICKEN

Serves 4-6

*12 chicken portions*
*10 ml salt*

**Marinade**

*250 ml olive oil*
*4 cloves garlic, crushed*
*65 ml lemon juice*
*10 ml fresh oregano*

Season the chicken portions with salt.

Combine the marinade ingredients well and marinate the chicken for 24 hours in the refrigerator, turning often.

Remove the chicken from the marinade and grill the portions over hot coals, basting often with the marinade, until cooked through - 20-30 minutes.

Serve with freshly baked bread and a Greek salad.

## SWEET-AND-SOUR GRILLED CHICKEN

Serves 6

*6 chicken thighs*
*6 chicken drumsticks*
*5 ml salt*
*2 ml freshly ground white pepper*

**Marinade**

*1 piece fresh root ginger, peeled and sliced*
*50 ml sunflower oil*
*30 ml honey*
*1 clove garlic, crushed*
*15 ml soy sauce*
*15 ml fresh lemon juice*
*15 ml dry sherry*

Season the chicken with the salt and pepper.

Combine all the marinade ingredients and marinate the chicken for at least 2 hours.

Remove the chicken from the marinade and braai over the coals, turning and basting often with the marinade, until golden and cooked through - about 30 minutes.

Serve with fried rice and stir-fried vegetables.

*Notes:*

Chopped preserved ginger or ground ginger may be used instead of fresh. If using ground ginger, sprinkle it on the chicken portions.

The stir-fried vegetables may also be cooked over the coals. Heat a heavy cast-iron frying pan, add a little sunflower oil and stir-fry the vegetables of your choice until cooked but still crunchy - 3-4 minutes.

## CHICKEN SOSATIES WITH LEMON TARRAGON MARINADE

Serves 4

*250 g boned chicken, cut into large cubes*  
*salt and freshly ground black pepper*

**Lemon tarragon marinade**

*250 ml olive oil*  
*75 ml fresh lemon juice*  
*3 spring onions, thinly sliced*  
*10 ml dried tarragon*  
*10 ml Dijon mustard*  
*2 ml salt*  
*1 ml freshly ground black pepper*

Thread the chicken cubes onto wooden skewers and season them lightly with salt and pepper.

Combine the marinade ingredients and marinate the chicken for 3 hours, turning occasionally.

Remove the chicken from the marinade and braai over hot coals, turning and basting frequently with the marinade, until golden brown and cooked through - 20-25 minutes.

Serve with spicy rice and a mixed salad.

## TURKEY KEBABS

Serves 4

*500 g boneless turkey meat*  
*300 g thick bacon rashers*  
*8 large button mushrooms*  
*8 pickling onions*

**Basting sauce**

*100 ml sunflower oil*  
*juice of ½ lemon*  
*salt and freshly ground black pepper*  
*10 ml fresh thyme*

Cube the turkey meat and the bacon. Thread alternately onto 8 skewers with the mushrooms and onions.

Combine the basting sauce ingredients in a screw-topped jar and shake well.

*Right: Citrus-marinated grilled butterfly chicken (p. 109)*

Brush the meat and vegetables on the skewers with the basting sauce. Braai about 15 cm above the coals, turning and basting often, for 15-20 minutes, or until browned and cooked through.

Serve immediately with baked potatoes and a mixed salad.

*Notes:*

Any other uncooked poultry may be used instead of turkey.

Turkey kebabs may also be cooked under the grill.

## HERBED CHICKEN POTJIE

Serves 6-8

*30 ml sunflower oil*
*3 kg chicken portions*
*2 leeks, sliced*
*1 clove garlic, crushed*
*200 g new potatoes, well scrubbed*
*4 tomatoes, skinned and chopped*
*200 g mushrooms, sliced*
*1 green pepper, seeded and sliced*
*125 ml dry white wine*
*125 ml chicken stock**
*100 ml chopped fresh parsley*
*15 ml chopped fresh savory or oregano*

Heat the oil in a heavy cast-iron pot over the coals and fry the chicken in it, turning frequently, until browned – about 5 minutes.

Add the leeks and garlic and fry for 1 minute. Layer the potatoes, tomatoes, mushrooms and green pepper over the chicken and add the wine, stock and herbs.

Simmer gently, covered, over the coals for 1 hour.

Adjust the seasoning, if necessary, and serve with cooked cracked wheat or samp.

*Note:*

To reduce the quantity of liquid, cook the potjie, uncovered, for the last 10-15 minutes of cooking time.

*Left: Traditional chicken pie (p. 115)*

## PHEASANT POTJIE

Serves 8

*125 ml sunflower oil*
*30 ml butter*
*1 x 1,5 kg pheasant, cut into portions*
*flour seasoned with salt and pepper*
*250 g rindless bacon, chopped*
*1 large onion, sliced*
*2 small carrots, chopped*
*1 stalk celery, chopped*
*2 cloves garlic, crushed*
*250 ml chicken stock**
*500 ml dry red wine*
*2 sprigs parsley*
*2 bay leaves*
*1 sprig thyme*
*1 x 1,5 kg chicken, cut into portions*
*juice of 1 lemon*
*salt and freshly ground black pepper*
*20 baby onions*
*20 button mushrooms*
*20 new potatoes*
*200 g broccoli, in florets*

Heat the oil and butter in a cast-iron pot over the coals, coat the pheasant in seasoned flour and brown for 5 minutes on either side.

Add the bacon and sauté until golden, about 5 minutes.

Add the onion, carrots, celery and garlic and sauté until the onion is transparent, about 5 minutes.

Add the chicken stock, 250 ml wine, parsley, bay leaves and thyme and simmer, covered, for 40 minutes.

Meanwhile, grill the chicken pieces quickly over hot coals until well browned, 5-7 minutes, turning and basting often with oil and lemon juice. Add to the pot.

Add the remaining wine to the pot.

Adjust the seasoning and simmer, covered, for a further 10 minutes.

Add the baby onions, then layer the mushrooms, potatoes and broccoli on top. Simmer, covered, for 1 hour.

Serve with crusty pot bread or mealie bread and a salad.

## TRADITIONAL CHICKEN PIE

Serves 6

*250 g flaky pastry**  
*1 egg yolk, mixed with a little milk*

**Filling**

*1 x 2 kg chicken, cut into portions*  
*6 whole pimentoes*  
*6 peppercorns*  
*3 cloves*  
*6 blades mace*  
*2 onions, sliced*  
*10 ml salt*  
*200 ml chicken stock**  
*30 ml sago, soaked in water until soft*  
*150 ml dry white wine*  
*1 egg yolk*  
*100 ml fresh lemon juice*  
*2 hard-boiled eggs, sliced*

To make the filling, stew the chicken with the pimentoes, peppercorns, cloves, mace, onions, salt and stock until tender – 45 minutes to 1 hour. Remove the spices and bones and cut the chicken into smaller pieces.

Add the sago and wine and simmer until the sago is transparent, about 10 minutes. Remove from the stove and stir in the egg yolk and lemon juice.

Roll out the pastry and line a large pie dish with half. Spoon in the filling. Place a few slices hard-boiled egg on top and cover with the remaining pastry, crimping the edges together to seal them. Make vents in the top of the pastry and brush with egg yolk mixed with milk.

Bake at 200 °C for 15 minutes, reduce the heat to 180 °C and bake for a further 15-20 minutes or until golden.

Serve hot, with sambals and yellow rice with raisins.

## PHEASANT PIE

Serves 8

*250 g flaky or short crust pastry**
*10 ml milk*

**Filling**
*1 x 1,5 kg pheasant, cleaned and cut into portions*
*75 g pork belly fat*
*6 cloves*
*2 bay leaves*
*1 clove garlic, crushed*
*5 peppercorns*
*10 ml salt*
*5 ml chopped fresh mixed herbs*
*30 ml vinegar*
*100 ml sweet red wine*
*10 ml cake flour*
*15 ml fruit chutney*
*100 g smoked bacon or ham, cubed*

First make the filling. Place the pheasant, pork belly fat, spices and salt in a large saucepan and add enough water just to cover. Simmer, covered, over low heat for 4-5 hours, or cook in the oven at 150 °C. Add the herbs, vinegar and wine after 3 hours.

Remove the meat from the sauce with a slotted spoon and discard the bones. Cube the meat.

Strain the sauce and stir in the flour, then combine it with the meat, chutney, bacon or ham.

Roll out half the pastry and use it to line a pie dish. Pour in the filling, then roll out the remaining pastry and use it to cover the filling. Crimp the edges of the pastry to seal it, then cut a few vents in the top to allow steam to escape during baking. Brush with the milk.

Bake at 250 °C for 10 minutes, then reduce the temperature to 180 °C and bake for a further 25 minutes, or until the pastry is golden.

Serve the pie sliced, with steamed young green beans, toasted almonds and stewed quinces.

*Note:*
Guinea fowl makes an excellent filling for this pie.

## CHICKEN COTTAGE PIE

Serves 6

*750 ml finely chopped cooked chicken*
*200 ml chicken stock**
*30 ml chutney*
*2 hard-boiled eggs, chopped (optional)*
*10 ml butter or margarine*

**Potato topping**
*750 g potatoes, peeled*
*125 ml hot milk*
*45 ml butter or margarine*
*2 ml grated nutmeg*

First make the topping. Quarter the potatoes and boil them in salted water for 15-20 minutes, or until tender. Drain well, then mash the potatoes with the milk, butter or margarine and nutmeg.

Spoon the chicken into an ovenproof dish. Combine the stock with the chutney and eggs and fold into the chicken.

Spoon the mashed potatoes over the chicken, dot with butter or margarine and bake at 190 °C for 30 minutes.

Serve with lightly steamed vegetables.

*Notes:*

Leftover turkey may also be used – turkey combined with meat stuffing is delicious.

Use 1 x 410 g can cream of mushroom or cream of asparagus soup instead of the stock and chutney.

# THAT LITTLE BIT EXTRA

Roasted poultry tastes just that little bit better with a delectable stuffing; a fruity or spicy marinade enhances poultry to be grilled; and no baked dish would be worthy of its name without the right sauce. Although many of the recipes in this book have their own special stuffing, marinade or sauce, we've given a selection of extra recipes for you to try out yourselves – and perhaps discover a totally new taste sensation. You'll also find the recipes for the pastry we've used, as well as a recipe for batter and for giblet gravy – just in case you need them.

## GIBLET GRAVY

Makes 500 ml

*giblets (neck, gizzard, heart, liver) from chicken or duck*
*pan juices from roasted poultry*
*50 ml cake flour*
*15 ml finely chopped fresh parsley*
*5 ml salt*
*2 ml freshly ground black pepper*

Boil the neck, gizzard and heart in lightly salted water until tender – about 2 hours. Add the liver and boil a further 15 minutes. Drain and make up the stock to 500 ml with water.

Dice meat, discarding gristle and bone.

Heat the pan juices in a saucepan, stir in the flour and cook until foamy.

Gradually stir in the stock and cook, stirring, until the gravy thickens – about 3 minutes.

Stir in the diced giblets and parsley, and season with salt and pepper.

Serve with poultry.

## SATÉ MARINADE

Makes about 350 ml

*125 ml minced onion*
*65 ml peanut or sunflower oil*
*65 ml fresh lemon juice*
*45 ml soy sauce*
*30 ml dry sherry*
*15 ml sugar*
*15 ml minced root ginger*
*3 cloves garlic, crushed*
*5 ml red pepper flakes*
*5 ml salt*

Combine marinade ingredients well and marinate poultry for 24 hours, or use as a basting sauce.

## WHISKY MARINADE

Makes about 250 ml

*100 ml whisky*
*65 ml soy sauce*
*65 ml firmly packed brown sugar*
*1 large onion, chopped*
*45 ml Dijon mustard*
*5 ml Worcester sauce*

Combine all the ingredients and marinate poultry overnight, refrigerated. Use as a basting sauce while braaiing poultry.

## GINGER SOY MARINADE

Makes 375 ml

*125 ml soy sauce*
*100 ml honey*
*100 ml dry sherry*
*1 clove garlic, crushed*
*5 ml grated fresh root ginger*

Combine the marinade ingredients in a saucepan and bring to the boil.

Simmer for 5 minutes, stirring, and let the marinade cool in the saucepan.

Transfer to a ceramic or glass dish and marinate poultry for at least 4 hours. Use remaining marinade as a basting sauce while grilling or braaiing poultry.

## PLUM BASTING SAUCE

Makes 500 ml

*375 ml stoned, chopped ripe Santa Rosa or Satsuma plums*
*65 ml fresh lemon juice*
*45 ml brown sugar*
*30 ml melted butter*
*15 ml grated onion*
*10 ml Dijon mustard*
*2 ml ground cinnamon*
*salt*

Purée the plums and transfer to a stainless steel or enamel saucepan; add the lemon juice, brown sugar, butter, onion, mustard, cinnamon and salt to taste.

Bring to the boil, then simmer, stirring, for 10 minutes.

Use to baste grilled or braaied poultry.

*Note:*
This may also be served as a sauce with roast guinea fowl or pheasant.

## ROSEMARY BUTTER STUFFING

Makes enough for a 1,5-2 kg chicken

*100 g butter*
*500 ml fresh bread crumbs*
*1 small onion, finely chopped*
*1 egg*
*15 ml finely chopped fresh or 10 ml dried rosemary*
*2 ml salt*
*2 ml freshly ground black pepper*

Heat the butter in a frying pan and toss the bread crumbs in it until crisp.

Combine with all the other ingredients.

## PRUNE AND BACON STUFFING

Makes enough for a 2 kg pheasant or Muscovy duck

*6 rashers rindless bacon, chopped*
*1 small onion, finely chopped*
*250 ml stoned, soaked and finely chopped prunes*
*250 ml fresh white bread crumbs*
*30 ml chicken stock**
*5 ml chopped fresh or dried oregano*
*5 ml salt*
*2 ml freshly ground black pepper*

Fry the bacon until transparent – about 2 minutes. Add the onion and sauté for 2 minutes longer. Remove from the frying pan.

Combine with all the other ingredients.

## HERBED MUSHROOM STUFFING

Makes enough for a 2 kg guinea fowl, pheasant or turkey

*500 ml fresh white bread crumbs*
*125 g mushrooms, finely chopped*
*30 ml chopped fresh parsley*
*10 ml chopped fresh thyme*
*1 small onion, chopped*
*1 egg*
*30 ml melted butter or sunflower oil*

Combine all the ingredients well.

## CHESTNUT STUFFING

Makes enough for a 2 kg turkey or chicken

*turkey giblets*
*250 g minced pork*
*1 onion, chopped*
*150 g mushrooms, chopped*
*15 ml butter*
*250 g chestnuts, chopped*
*2 slices white bread, crusts removed*
*200 ml dry red wine*
*1 egg*
*5 ml chopped fresh thyme*
*5 ml salt*
*2 ml freshly ground black pepper*

Chop the giblets and combine with the minced pork.

Sauté the onion and mushrooms in the butter until softened, about 5 minutes, and combine with all the other ingredients.

## BREAD STUFFING

Makes enough for a 1,5 kg chicken

*30 ml butter or margarine*
*15 ml finely chopped onion*
*500 ml dry white bread crumbs*
*125 ml milk*
*5 ml finely chopped fresh parsley*
*2 ml dried thyme*
*2 ml celery salt*
*5 ml salt*
*pinch freshly ground black pepper*
*1 egg (optional)*

Melt the butter or margarine in a small frying pan and sauté the onion until transparent, about 5 minutes.

Add remaining ingredients, except the egg, and mix well. If mixture is too crumbly, bind it with the egg.

*Note:*
Make 1½ times the quantity to stuff a turkey.

## SAUSAGE AND APPLE STUFFING

Makes enough to stuff a 2-2,5 kg Muscovy duck or chicken

*250 g pork sausage meat*
*15 ml butter*
*1 onion, chopped*
*1 stalk celery, chopped*
*1 large Granny Smith apple, peeled, cored and cubed*
*1 thick slice white bread, crusts removed*
*1 egg*
*5 ml salt*
*2 ml freshly ground black pepper*

Sauté the sausage meat in the melted butter for 10 minutes. Remove from the frying pan.

Add the onion, celery and apple to the pan and sauté 7-10 minutes. Remove from the frying pan.

Combine with the sausage meat and the remaining ingredients.

## CITRUS RAISIN STUFFING

Makes enough for a 2 kg turkey or goose

*1 ℓ fresh white bread crumbs*
*2 ml salt*
*2 ml freshly ground black pepper*
*finely grated rind of 1 lemon*
*finely grated rind and juice of 1 orange*
*50 ml butter or margarine, cubed*
*85 g seedless raisins or sultanas*
*50 ml finely chopped fresh parsley*
*2 eggs, beaten*

Combine bread crumbs, salt, pepper, lemon and orange rind in a bowl. Add butter or margarine and rub in until well distributed. Stir in raisins or sultanas and parsley and use orange juice and eggs to bind the mixture.

## FRUITY NUT STUFFING

Makes enough for a 2 kg turkey

125 g *dried apricots, soaked in sherry until plump*
45 ml *butter*
1 *onion, finely chopped*
200 g *pine kernels or almonds, finely chopped*
2 *stalks celery, finely chopped*
375 ml *fresh bread crumbs*
30 ml *finely chopped fresh parsley*
5 ml *salt*
2 ml *freshly ground black pepper*

Drain the apricots and chop them finely.

Melt the butter in a saucepan and sauté the onion until transparent, about 5 minutes.

Stir in the remaining ingredients and leave to cool.

## BÉCHAMEL SAUCE

Makes 250 ml

1 *small carrot*
½ *onion*
1 *stalk celery*
1 *blade mace*
2 *peppercorns*
250 ml *milk*
30 ml *butter or margarine*
60 ml *cake flour*
2 ml *salt*

Place the carrot, onion, celery, mace, peppercorns and milk in a saucepan and bring milk to simmering point over low heat. Remove from stove and leave, covered, for 30 minutes.

Strain the liquid through a sieve and discard the vegetables.

In another saucepan, melt the butter or margarine over low heat and add the flour, stirring constantly. Cook for 2 minutes, stirring, then remove from the stove. Gradually add the flavoured milk, stirring constantly and season.

Return to the stove and simmer for 5 minutes, or until thickened, stirring constantly.

*Note:*
For a thick sauce, use 50 ml butter and 100 ml flour.

## WHITE SAUCE

Makes 250 ml

*Prepare as for Béchamel sauce, using plain instead of flavoured milk.*

## CHEESE SAUCE

Makes 250 ml

*Prepare as for Béchamel or white sauce, adding 50 g mature Cheddar cheese, grated, just before serving.*

## ASPARAGUS SAUCE

Makes 400 ml

*30 ml butter*
*30 ml cake flour*
*1 x 410 g can asparagus pieces*
*250 ml milk*
*1 egg yolk*
*125 ml cream*
*5 ml salt*
*2 ml freshly ground black pepper*

Heat the butter in a saucepan and stir in the flour to make a paste (roux).

Drain the asparagus and reserve the canning liquid.

Add the milk and 125 ml reserved liquid and simmer, stirring constantly, for about 10 minutes or until smooth and creamy.

Beat the egg yolk with the cream. Remove from the stove and add a little of the hot sauce to the egg mixture. Add to the sauce, season with salt and pepper.

Serve with roast chicken.

## TANGY MUSTARD SAUCE

Makes 300 ml

*75 g sugar*
*4 eggs*
*30 ml mustard powder*
*15 ml cake flour*
*5 ml salt*
*2 ml freshly ground black pepper*
*250 ml red-wine vinegar*

Beat 60 g of the sugar and the eggs together and pour into a saucepan.

Combine the remaining sugar, mustard powder, flour, salt and pepper. Add the vinegar and stir into the egg mixture in the saucepan. Heat slowly, stirring constantly, and remove from stove as soon as the sauce begins to thicken.

Serve with any roast or smoked poultry.

*Note:*
This sauce makes an excellent dip. It will keep for up to 1 week in the refrigerator.

## PEANUT SAUCE

Makes 375 ml

*Saté marinade**
*45 ml smooth peanut butter*
*15 ml soy sauce*
*250 ml thick cream*

Purée the marinade, peanut butter and soy sauce in a blender or food processor.

With the motor running, add the cream in a steady stream until well blended.

Transfer the sauce to a stainless-steel or enamel saucepan and simmer until slightly thickened, about 10 minutes. Serve hot.

## BATTER FOR FRIED CHICKEN

Makes enough for 2 kg chicken portions

*250 ml cake flour*
*2 ml salt*
*2 eggs, separated*
*45 ml melted butter*
*200 ml beer or water*

Combine the flour, salt, egg yolks and butter in a bowl.

Gradually beat in the beer or water and leave to stand for 1 hour.

Whisk the egg whites and fold them into the batter just before use.

Dip the chicken portions in the batter and either shallow or deep-fry them until crisp.

## QUICK FLAKY PASTRY

Makes about 1 kg

*500 g cake flour*
*7 ml cream of tartar*
*5 ml salt*
*500 g well-chilled butter, grated*
*15 ml fresh lemon juice or brandy*
*150 ml ice water*

Sift flour, cream of tartar and salt into a mixing bowl. Cut the butter into the flour.

Combine the lemon juice or brandy with the water and add to the flour mixture.

Knead lightly into a ball and refrigerate for 15 minutes.

Roll dough out thinly on a lightly floured surface. Fold over and roll out again. Repeat.

Refrigerate for up to 1 day, or freeze for up to 3 months.

## SHORT CRUST PASTRY

Makes 1 kg

*500 g cake flour*
*5 ml salt*
*5 ml baking powder*
*350 g chilled butter, grated*
*175 ml ice water*
*25 ml fresh lemon juice*

Sift the dry ingredients together into a bowl. Rub the butter into the mixture until it resembles coarse crumbs.

Sprinkle water and lemon juice on the crumbs and combine lightly. Wrap in waxproof paper and chill in the refrigerator for up to 1 day or freeze for up to 3 months.

*Note:*
For a richer pastry, use 150 ml ice water and 1 egg yolk instead of 175 ml ice water.

# INDEX OF RECIPES